Television News

Fourth Edition

Ivor Yorke

Revised by Ray Alexander

Focal Press

OXFORD AUCKLAND BOSTON JOHANNESBURG MELBOURNE NEW DELHI

Focal Press
An imprint of Butterworth-Heinemann
Linacre House, Jordan Hill, Oxford OX2 8DP
225 Wildwood Avenue, Woburn, MA 01801-2041
A division of Reed Educational and Professional Publishing Ltd

 A member of the Reed Elsevier plc group

First published as *The Technique of Television News* 1978
Second edition 1987, reprinted 1990
Third edition 1995
Fourth edition 2000

British Library Cataloguing in Publication Data
Yorke, Ivor
Television news. – 4th ed.
1. Television broadcasting of news
I. Title II. Alexander, Ray
070.1'95

Library of Congress Cataloguing in Publication Data
A catalogue record for this book is available from the Library of Congress

ISBN 0 240 51615 X

Composition by Scribe Design, Gillingham, Kent
Printed in Great Britain

Contents

About the author

Ivor Yorke was a journalist on local newspapers and in Fleet Street before joining BBC Television News as a sub-editor in 1964. He went on to become a reporter, producer and editor on a wide range of BBC news programmes. After 28 years in the BBC, the last six as Head of Journalist Training, he left to establish Broadcast Techniques, his own training consultancy, working for professional broadcasting organizations in the United Kingdom and abroad, and teaching television journalism at several colleges. From 1992 Ivor Yorke was vice-chairman of the National Council for the Training of Broadcast Journalists and chaired the industry working party establishing national vocational qualifications for broadcast journalism. Sadly, Ivor died in 1996.

About the revising author

Ray Alexander studied journalism and law, and worked first as a reporter for the *Daily Telegraph* in Belfast and on regional newspapers in England. He started in broadcasting as a producer at BBC Radio Merseyside, moving on to BBC Television's London Newsroom, BBC Television in Northern Ireland and BBC Radio in London as a producer and reporter. He was Senior Political Correspondent and Diplomatic Correspondent for *TVam*, and *TVam Frost On Sunday*. He has two qualifications in instructional techniques and is a director of the UK Broadcast Journalism Training Council. He is manager and senior trainer for BBC news training.

Acknowledgements

Acknowledgements to third edition

My grateful thanks especially to Ed Boyce at Meridian Broadcasting, and Tom Evans at ITN for sparing time away from their own duties – Ed as Head of Training and Tom as a programme editor – to cast their eyes over the finished manuscript and to make some very helpful comments, many of which I have incorporated.

Thanks too to Nik Gowing, Diplomatic Editor at Channel 4 News, for allowing me to quote from his excellent paper on the effects – or otherwise – of foreign news reporting on the policy of governments, with particular emphasis on the long-drawn-out conflict in Bosnia; to Ali Rashid, editor of Calendar, and Martin Brooks, Roger Bufton, Andy Griffee, Tim Manning, Gordon Macmillan, Roger Protheroe, and Roy Saatchi of the BBC for giving me an informal insight into what they look for in potential recruits to television news; Neil Everton for his story of multi-skilling success in Canada; Miriam Darby of Ceefax, Gail Gillogaley, of Teletext Ltd and Rogger Oppé of VGB for bringing me up to date on teletext and for the use of photographs; Avon Harpley, Hat Trick Productions, Anne Darwent of Channel 4 Stills and Tim Hill for allowing me to use an illustration from *Drop the Dead Donkey*; William Hood for pictures of Newsforce in action; Lynne Gardiner, Avid Technology; George Maher, Plain English Campaign; Bill Meredith (BARB), Andrew Hemming, BBC Audience Research; Ian Mizen and Stephen J. Steele of Vinten; Jim Keating, EDS Portaprompt; Jean-Pierre Julien of the EBU; Charlie Raynsford of Newstar; Susie Spragg of ITN; David Robinson, International Weather Productions and Gail Yamakazi, Intelsat.

IY, 1995

Acknowledgements to fourth edition

In this fourth edition I would like to include additional acknowledgements to Sky Television; the BBC; Ananova Ltd; Wordsearch/Alan Williams; Quantel Limited; and Gary Braid for figures they have kindly supplied.

RA, 2000

1

Introduction

This book is for people who want to work in television news, are interested in the practical techniques of television news, or who want to know how it fits into the multimedia/Net-media make-your-own-news world we now live in.

To understand fully what the place of television news can be in the first decade of the new century we also need to know how it developed. There were three ages of broadcasting. First, wireless telegraphy and radio. They revolutionized commerce and communications. Among thousands of historical moments our grandparents may have known was Neville Chamberlain's announcement that Britain was at war with Germany. Or, in America, the day when the actor Orson Welles terrified the audience with his theatre group's performance of *The War of the Worlds*. A work of fiction appeared to be reality – not just because of the way the drama was presented as 'live news' (Earth is being invaded, right now, by scary Martians!) but because of the mood and times in which people lived their real lives.

Then the second wave: television. Many business people did not regard television as having any commercial potential. Like the telephone, there was initially nobody to talk to, or broadcast to. This *device*, as its critics described it, would be no good because the word was half-Greek and half-Latin. The first event filmed specifically for television news was by the BBC in August 1936, when the ocean liner, *Queen Mary*, docked in Southampton. The pictures were transmitted four hours later from the home of the new BBC Television News service at Alexandra Palace, a Victorian pile situated in parkland in north London. Four hours to make transmission was a major achievement and this was long before a fully-staffed television newsroom had been established. That was followed in September 1936 by the German service, Reich Rundfunkgesellschaft, which filmed the Nuremberg Rally, flew the material to Berlin, and transmitted it the same evening. The race against time had begun. News would never be the same again.

Now, in the third age, there is television news in which the viewer can choose to select pre-prepared news items, or select only news services which are continuous and often live. The race against time is irrelevant, because 'we are live'. It may be live for no particular reason, but you can see events happen as they happen. There is also television news linked to the Internet, with news discussion groups in which anyone can have their say about news events – all this, plus digital or interactive services, widescreen viewing and home shopping. Twenty-four hour news, years ago the monopoly of CNN, is now on hundreds of cable services. The problem that faces television news journalism is convincing the audience that self-selection of news subjects, news-that-suits-me, will mean that vast numbers of people may not be aware of the real world at all. They can airbrush out the unpleasant and unwanted events of life. People do not always know what they need to know. With self-selection there could be millions of people who would never have known about a war in Sierra Leone, let alone been able to place it on a map. If they do not know about such things, then why should they ever care.

If we sit back after consuming this mass-menu of television news we should ponder the future by remembering how it came about. On a summer Monday in 1994, in an elegant function room at Alexandra Palace, about a hundred men and women began to assemble to celebrate a special anniversary. Forty years earlier, in a now disused, dilapidated studio at the other end of the building, a handful of the veterans among these partygoers had been midwives at the birth of the first regular television news service in the United Kingdom. The pathfinders who had made television news were retired by the time it matured into the modern digital services we now take for granted. What the 1994 reunion group had in common with the pioneers of Internet services in that year was recognizing a moment in history when they were doing something new, that they were changing the world of communication and knowledge.

At Alexandra Palace, the speeches, the brief video compilation of highlights and the conversations with old friends and colleagues over a convivial glass or two must have stirred marvellous memories, but it is doubtful whether any of those involved in the first edition of the BBC's *News and Newsreel* broadcast at 7.30 on the evening of 5 July 1954, could possibly have imagined at the time that one day television news would supersede print and radio as the most powerful and effective form of journalism. Or that television news would bond with cable, satellites and computer-based technology to find itself spread across the globe and then onto computer screens at home. The presentation of that first evening's lead story in 1954, about continuing truce talks in Indo-China, would hardly have seemed to hold the promise of such a glittering and controversial future: a caption title reminiscent of cinema newsreels, a map and a couple of agency photographs, accompanied by a sparse voice-over commentary and discreet music.

The evolution of television news, 1954–2000

The ability of television news to influence public opinion to any significant extent was probably not fully recognized until the mid-1960s, after the broadcasters had demonstrated that new communications technology, combined with a willingness among some services to cooperate regularly in the exchange of news material, could make pictures of any important event available beyond national boundaries within hours. Over the years world audiences shared the John F. Kennedy assassination, student riots, Watergate, terrorism and various wars including Vietnam and the Middle East, and nothing could ever be the same again.

So, by the 1980s, anyone who remained sceptical about the power of television news to move public opinion must have had all doubts swept aside by the astonishing, spontaneous response to the appearance in October 1984 of harrowing pictures of famine in Ethiopia. The impetus for the creation of the Band Aid relief fund and all that has followed in an attempt to alleviate the suffering of millions can be attributed directly to the reports seen on the news bulletins of an estimated 400-plus broadcasting organizations. The same pool of material from television news influenced opinion about what was happening in the late 1990s in Rwanda, Kosovo and Indonesia.

Today's news programming has become an accepted part of the culture of every society which embraces television. Those who report and present the news are famous enough to be caricatured. Their faces adorn magazine front pages. Their on-screen performances, and the salaries they earn for them, are the subject of endless discussion and speculation. How they live their private lives, what they wear, what they do and what they say (especially if it is controversial in the slightest) are followed with almost indecent interest by press and public.

At last there is fiction, too, about television journalism – and American films, although neither *Network* nor *Broadcast News* ranks with an old classic, *The Front Page*. It is also seen as worthy of parody – by television. *Drop the Dead Donkey*, a 1990s' situation comedy series set in a television newsroom and *The Day Today*, a spoof news programme broadcast opposite the BBC's *Nine O'Clock News*, achieved cult status – even if most of the cult's adherents were themselves journalists searching for clues to the real identities on whom the fictional characters were based.

Television news has become a respectable subject for serious academic study, especially by sociologists postulating theories about the role and motivation of the practitioners, and the influence they bring to bear on 'agenda-setting' – what stories they choose to cover and then how they process them.

Every so often an aggrieved public figure will complain angrily about stories 'got up by the media' or the concentration on personalities rather than issues. The media were

Figure 1.1 The face of television news for the 21st Century. Ananova, the newsreader at www.ananova.com website, is a fully animated cyber character who may well threaten the large salaries of human newsreaders. (Courtesy of Ananova Ltd, formerly PA New Media)

accused of being responsible for stirring up stories about perils for society likely to be caused by the 'Millennium Bug', when of course 'the media' could have ignored all those computer scientists gushing with anxiety and doomsday warnings about computer programs which relied upon clock-based chips. There is in British television news history, so far, only one occasion when the criticism was fundamental enough to warrant being met head on. In the autumn of 1986, BBC Television News executives took the unprecedented step of issuing a line-by-line rebuttal of charges made by the then ruling Conservative Party. The complaints, detailed to the point of challenging certain words used in the headlines, were about the way the BBC *Nine O'Clock News* had covered a controversial American bombing raid on Libya six months earlier. Overall public assessment of this bout of linen-washing was that if the politicians lost the argument they succeeded in making the journalists justify their actions to the audience to an extent previously unknown.

The suspicion exists that 'the establishment' – no matter what party holds power, or who runs the Church, the judiciary or the military/industrial complex – dictates what and how television reports. But there is probably greater genuine surprise that television journalists do not see it as their first duty to protect society from the unpalatable, the outwardly reasonable view being that the world would be a better place if items about civil disturbances and similarly distasteful events were simply not shown. There exists a belief that those who are responsible for television news programmes demonstrate their partiality by the very act of reporting the existence of dissent, and a theory that various royal marriages would still be intact and several politicians still in their posts if television had not made matters worse by taking up, however reluctantly at first, stories based on the newspaper agenda.

The paraphernalia of television itself is considered to be provocative at times: in the days when the coverage of protest demonstrations featured regularly in diaries for otherwise quiet weekend news bulletins, it took a while for the journalists to spot the coincidence between the outbreak of trouble and the appearance of the camera. Suggestions of a 'copycat' phenomenon were raised about a series of riots in British cities in the early and late 1980s. It happened again when riots broke out in Britain and America around the time of the trade talks in Seattle in late 1999. The demonstrators appear to have been highly offended by the suggestion that they were influenced by what they saw on television news, and chiefly tried to convey the message that it was the police who were to blame. 'The police themselves dismissed the idea that the violence was copied from television. They saw the causes of the riots as many and varied: social deprivation, frustration, and hate.'[1]

Governments seem to have no doubt about the influence of television. Most democratic governments even build the reaction of the public to television news reports into their foreign and diplomatic policy – What will people say if the TV news reports that we bombed civilians?

The coexistence between the political/military establishment and world television journalism was mostly laid down during the original Gulf War, fought out in Kuwait and Iraq early in 1991. Months of diplomatic shadow-boxing before the fighting gave the military time to draw up ground rules. Journalists were attached to Media Response Teams – pools – under the eye of a military officer who viewed reports before they were transmitted to ensure that Ministry of Defence guidelines were not broken. These forbade among other things any references to the number and location of troops and weapons and future battle plans. No details of dead soldiers were to be given until their

1. Howard Tumber (1992). *Television and the Riots*, British Film Institute.

families had been told. A few brave – or difficult – souls doing things on their own initiative were arrested, while about 300 other journalists threatened to get in their cars and drive to the front unless the US-led coalition gave them more access to the combat zones.

For television the main significance of the Gulf War was its use of satellite technology. For the first time an international news event was dominated by *live* television coverage – more specifically by CNN, the 24-hour news channel, which relayed events unedited to a global audience. Reports of the first raid on Baghdad and Iraqi missile attacks on Israel made particularly riveting viewing. At times it seemed that even the politicians in the White House and Baghdad were relying on CNN for their information. CNN's performance was not universally acclaimed, and the network's relationship with the Iraqi authorities was questioned, particularly in the period when all other Western news media had been expelled. There were other worries, too: 'live coverage of the damage caused by Scud missile attacks on Israeli cities could enable the enemy to identify the random sites on which their projectiles had fallen and to readjust their targeting'.[2]

But the Gulf War left a permanent mark. More players have hastened to join the ranks of the all-news leaguers, as personified by CNN, Sky News, BBC World, BBC News 24 and others, and the portable satellite-uplink has become an indispensable tool in the reporting of events. Without it, it is doubtful whether many of the other big stories attracting huge and continuing coverage in the mid-1990s, among them the internecine conflicts in Bosnia and Rwanda, would have made such an impact.

As Nik Gowing, a television journalist with long experience in covering diplomatic events for ITN and the BBC, puts it bluntly in a fascinating study of the impact of television's coverage of armed conflicts and diplomatic crises: 'Often no dish means no coverage.'[3] What Gowing defines as 'real-time television' – live or very recently recorded pictures beamed back by satellite – has created a new grammar and editorial agenda for TV news coverage, bringing with it fresh dynamics and pressures. But his paper, based on his own experiences and more than a hundred interviews with senior officials and politicians at the heart of policy-making in several countries, challenges the belief that television's ability to provide instant 'video ticker-tape' should be mistaken for a power to influence foreign policy.

In a study dominated by the appalling events in Bosnia, Gowing recognizes that 'on a few occasions it [the impact of television coverage] can be great, especially when it comes to responding with humanitarian aid. Routinely, however, there is little or no policy impact when the television pictures cry out for a determined, pro-active foreign policy response to end a conflict.'[4]

How television's insatiable appetite will continue to be filled is another matter, and whether viewers will be satisfied with analysis and interpretation (what others might call 'waffle') in the absence of real news is hard to gauge. The signs are that journalists will become more resistant to some of the more obvious attempts at news management and blandishments of the spin doctors. Questions are starting to be asked about the value of some of the international 'spectaculars' involving world politicians and statesmen, and the attitudes of those in authority who are happy to welcome the cameras when it suits them, but who seek to keep them out when matters of real importance occur.

2. Leader column, *The Independent*, 25 January 1991.
3. Nik Gowing (1994). Real-time television coverage of armed conflicts and diplomatic crises: does it pressure or distort foreign policy decisions?, paper, Harvard College.
4. *Ibid.*

A former editor of *The Sunday Times*, Harold Evans, who later spent many years working in journalism in America, expressed anxiety about the need for live and continuous news services to fill airtime. He described it as the 'tabloidisation' of news, notably on television that was available round-the-clock on several cable stations. 'They have discovered that the only way they can keep an audience is to hit a single story with everything they have got. OJ Simpson, Monica Lewinsky, Princess Diana, Kosovo, school shootings ... when stories like these are running, nothing else in the world is happening.'[5]

The mirror image of that stance can be seen in the large number of countries where reporting is not 'free' in any sense. Given that television in many countries is controlled or funded by governments, it is not difficult to appreciate that many news services are able to produce little except what is officially sanctioned. In addition, foreign camera crews and reporters cannot fly in with their equipment to anywhere they please and expect to start work. Some countries simply refuse to allow them entry: visas to get into others may take months, and when permission is finally granted the presence of 'minders' may be so inhibiting that the reports which are made may be no more informative than those old-fashioned cinema travelogues.

In the meantime, one effect is that some events about which there is no independent confirmation might just as well not have happened, for while pictures of, say, a serious accident on a foggy German autobahn would be available to the rest of the world in next to no time as a matter of routine, some natural disaster which wiped out a remote African village might go unreported because no cameras were there until months afterwards. In this way, even in the satellite age – just as no dish means no coverage of armed conflict – it is still possible for whole areas of the world to remain ignorant of what is taking place in others, either by reason of geographical accident or through the deliberate actions of governments anxious not to let any event of an unfavourable nature be seen by outsiders. As an insurance, among the first things some governments or their agents will do when contemplating something nasty will be to kill, intimidate, imprison or expel any journalists considered likely to be an embarrassment to them. Sometimes it is done quite blatantly.

The newspaper world, meanwhile, has maintained its morbid interest in the fortunes of television and television news, developing an ambivalent relationship which allows its reporters to dig deeply for the dirt while simultaneously devoting pages to programme previews, reviews and personality interviews. For although some journalists writing about television news display a distressing lack of understanding of the subject, they are usually cute enough to realize that television news is news.

Despite their own new technology and their lavish investment in their websites, newspapers are unable to compete for speed, have long been relegated to second place behind the electronic media in the coverage of news, and so have had to develop along different lines – the brash, irreverent tabloids and the weighty broadsheets offering the main contrasts in tone and style. Few papers are spared the need to battle for circulation, with cuts in the cover price, reader competitions and various special offers among the weapons employed.

Mass communication, like all forms of technology, evolves from one medium to the next. They need each other and the changes in mass communication do not jump abruptly, but overlap into stages. At the beginning of the twentieth century it was predicted that first radio, and then television, would put newspapers out of business. Then at the end of the century the Internet and the web were expected to put them *all*

5. *The Guardian*, 8 November 1999.

out of business, at least from the comments of people you would expect to say that. It is more likely that the web will force significant change on television news during this decade, but it is less likely that the actual content – assessed, judged and provided by humans – can be replaced. 'Computers can be programmed to play chess like Gary Kasparov but they can never paint like Picasso. They can mimic the look but never the touch, the passion, the emotion. These are inherent components of a human newsroom, not a machine.'[6]

The fact remains that very few detailed studies have been made into the problems, processes and techniques involved in bringing news to the screen 365 days a year. Admittedly it is not an easy subject to explain. Television news well done is not simply radio with pictures or identical to any other part of the television business. It has been described as a kind of electronic jigsaw puzzle which, like other puzzles, makes no sense until it has been completed. Taking a few separate pieces at random is rather like examining the big toe and thumb and expecting them to give an accurate picture of what the entire human body looks like. In television news the most important parts of the jigsaw are people, operating within their own spheres of activity along parallel lines which converge only at the times of transmission.

How those pieces fit together may not greatly interest the professional critics, but it does concern those journalists for whom television companies all over the world are continuing to provide worthwhile and stimulating employment. It also concerns the companies who, fully aware of the ever-mounting pressure upon them to produce news programmes of a professionally high standard, need the journalists they engage to demonstrate immediate expertise and reliability. The craft of writing clearly and succinctly is only part of it. Much of the rest of the battle has to do with the bewildering battery of electronic equipment and strange jargon which needs to be mastered properly before a single word is spoken. This means that journalists must understand the nature of television, what they are doing, why, and how their own roles fit in with the rest of the team.

More than ever before, all this has to be achieved in the full glare of the public spotlight, for while members of the viewing audience may expect the daily newspapers they buy to reflect their own political viewpoints, they have every right to demand that those involved in television news take seriously their responsibilities for accuracy, fairness and decency. Sadly, the fun days of television news went many years ago: too much is now at stake. So while it may be considered a denial of personal freedom, it is not too much to ask any television journalist voluntarily to stay free from active and formal allegiance to any political party or sensitive cause. As for 'committed' journalism of any hue, that has always seemed to me to be counter-productive, because one part of the audience is inevitably going to be alienated.

For all these reasons and more, every journalist planning a career in television must be attuned to the demands of an exacting occupation. That means a certain amount of formal training, which has become more important than ever as television techniques advance and which, significantly, seems to escape the severest cuts when economies are demanded in other areas. Fortunately, too, only a few dinosaurs remain to be convinced that training is a valuable way of equipping the recipient of it with the know-how and confidence to employ the marvellous technical resources which are about to become the tools of his or her trade.

The experienced newcomer to television news might have been employed on a radio station, news agency, magazine or a weekly or daily newspaper. The recruit from a

6. Piers Morgan, editor of *The Mirror*, writing in *The Guardian*, 6 December 1999.

journalism course will have a grasp of the fundamentals. The novice might be straight out of university or elsewhere and needs initiation into journalistic principles as well as the special skills needed for television. All have the same thing in common: they need to learn, preferably as swiftly as possible, how.

What follows in this book is not intended as a substitute for a properly run training scheme led by expert professional trainers. It makes no attempt to standardize editorial procedures or the processes guiding news selection and judgement because none is universally accepted. Neither does it claim to be comprehensive, as working practices and conditions, terminology, systems and equipment in use are bound to vary considerably according to the importance and financial resources of the station concerned.

What this book is intended to be is a first reader in basic television news style and production. In the face of the fast changes in the way real information and fact, together with individual comment and often unreliable rumour, are being moved, accessed and traded in the first decade of the new century, it is still essential for journalists or aspiring journalists to know about how television works at a practical level, together with enough background to put it into the modern context. It is an attempt to help newcomers fit the pieces of the electronic jigsaw puzzle together in as painless and non-technical a way as possible.

2

Getting into television

The popularity of broadcast journalism as a career remains undiminished despite the growth of Internet services and online news. Almost every halfway literate youngster, it seems, wants to be in 'the media'. If that's not being a part of the web, then it means television and ideally on screen. Television is seen as a glamorous occupation, with the dual prospect of fame and salaries in telephone numbers for an increasingly youthful band of personalities. Even many of those who start in radio and come to like the sound of their own voice develop a hankering for the visual medium, and to 'graduate' from the older discipline to the newer is seen as a natural career progression. To say television simply is more influential is not to downgrade the importance of radio, which has its own continuing attraction as a pure and spontaneous medium for information.

Not that an increase in the competition 'out there' is the only factor to bear in mind in any effort to succeed. Getting started in television for the beginner remains as hard as ever, despite the swing of the pendulum in the industry which means that some services contract as others expand. There was a flurry of opportunity in the late 1990s with the launch of Channel 5 and the BBC's 24-hour domestic television news service, News 24. Sky News was always there and always alert as a UK and international competitor to the BBC, CNN and ITN. There were also jobs at Bloomberg Television which established itself as a growing force in television. Agencies such as Reuters, and companies like GMTV, also remained fertile places for journalists to gain regular work. On the other hand there were takeovers within the terrestrial commercial television companies and plenty of local cable news services came and went within a short period. This led to some rationalization and a continuing trend towards casual shift work for journalists. Freelance working and short-term engagements – i.e. contracts of 364 days or below – became the norm.

The outcome is a limited market at entry level and a premium on experience. Those coming into television are frequently people with backgrounds in other forms of journalism. They already have the basic skills to offer and need only to be introduced to the intricacies of television before they can be contributing to programme output.

What qualities for the beginner?

Those undeterred by this scenario and still intent on making their way in a difficult, precarious, notoriously unsocial (and not always well-paid) occupation should first of all be aware of the qualities they need if they are to have any chance of getting beyond the front door.

A good education

Broadcast journalism is not an exclusively graduate occupation, but it does still help. That is because the completion of a degree course suggests the possession of an ordered mind and a developed intellect. (A good school record plus a few years on a local newspaper, with some good cuttings to show for it, counts for just as much, though.) When it comes to higher education, the question of the subject then arises. Any specialist course of study which includes the word 'broadcast', 'journalism', 'communications', or 'media' in the title seems to attract sufficient numbers of students to make it viable. Some, though promising, are more abstract than practical in nature. In general, students should make a judgement of a degree course on the level of practical skills included. You will be of more interest to a first employer if you know how to use a microphone. It is of little use if your degree was dominated by a study on the social implications of docu-soaps.

The often-expressed concern of employers is that some courses are more concerned with the theory and sociology of broadcast journalism rather than with the practicalities. The result is that some students come away with ideas which, though laudable in their way, are irrelevant to the grind of daily news programming. Employers considering the academic background of novices are inclined to give as much weight to the proven educational virtues of economics, politics, history, geography and – increasingly – science and a language, or indeed of any subject likely to be of practical use to a journalist. First-class honours are no guarantee of employment.

An insatiable interest in news and current events

Proving tenacity and curiousity are all. The burning desire to 'find out' what is going on and to tell as many people about it as possible is something the journalist never loses. Deep knowledge of and interest in one specialist subject is helpful, but at this stage not nearly enough. What potential employers are looking for is a good all-round awareness of current events, domestic and foreign. A passion for wanting to know what's going on and why will really impress an employer. Healthy scepticism is another virtue in every journalist, but while it is all very well to have strong political opinions and a highly developed social conscience, remember that as a television journalist doing a professional job you will have a duty to be impartial.

Michael Buerk, one of Britain's leading television journalists, is on record as being quite clear about it: 'There is a difference between being a reporter and a campaigner.'[1]

Evidence of a commitment to journalism

Some evidence of a genuine commitment to journalism is essential. Contributions to school or college magazines, involvement in college, hospital or community broadcasting, voluntary or otherwise, would be regarded as the minimum indication of real interest. But the experience does need to be relevant. A couple of terms reviewing plays or

1. *John Dunn Show*, BBC Radio 2, September 1994.

being a disc jockey are not likely to suggest a lifetime's ambition to become an investigative journalist.

The right personality

Journalists come in both sexes, all ages, shapes and sizes, and with many different temperaments. But whether you are inherently outward-going and happiest in company or a quiet loner, you are, in television, inescapably a small part of a machine. Any television news or current affairs broadcast represents the fusion of highly developed production and technical as well as editorial skills, and unless you have the ability to recognize the importance of teamwork, any other admirable qualities you might possess will not be enough.

'Broadcastability'

Not every broadcast journalist finds him or herself in front of the microphone or camera, and many organizations have established clear dividing lines between journalists who 'perform' (reporters, correspondents and presenters) and those who do not (editors, producers and writers). That is definitely changing, and although not being a broadcaster need be no bar to success in 'backroom' and managerial jobs, there is no doubt that an acceptable voice and appearance will improve opportunity.

Other skills

The ability to operate a keyboard and mouse will be taken for granted. Journalists are increasingly being expected to train in the operation of other equipment, including sound and vision desktop editing and digital portable cameras. Any technophobia is likely to be a drawback. Possession of a full, clean driving licence is assumed (but see the section 'Catering for special needs' later in this chapter).

The ability to write

Finally, you must be able to write. That's why you want to do the job! You might have great personal style, a great voice and loads of personal charm, but if you cannot write – you'll be rumbled within a week! Signs of natural ability are easy enough to spot, but what employers want is an indication of a journalistic approach, and they will not be convinced by examples of notes, journals or academic papers.

Eyes on your target

The first piece of advice for any would-be broadcast journalist is to have an idea of what you want to do, and while no one would expect you to be able to reel off exact titles or categories, a general desire to be 'in/on television' is not enough. Examine your own instincts. Are you by nature inclined to be a 'fire-engine chaser'? If so, you are more likely to be interested in daily hard news than the deeper, investigative journalism associated with documentary programmes. And if you are interested in documentaries, then remember not all fall into the category of news/current affairs. They may be part of scientific, historical or other strands produced by different programme areas.

Once you have an idea of the kind of programme which interests you, the next step is to decide what part you would like to play in it, given the opportunity, and to focus your efforts accordingly. But at this stage it is also essential to be realistic. If your ultimate goal is to be a foreign correspondent, don't expect to achieve it at once. Indeed chances are there will always be better, more talented people ahead of you. Nevertheless it is essential not to be deflected or sidetracked into an entirely unsuitable area from which it would be difficult to move.

Among the poorest advice given, in particular to young women wishing to become journalists, has been to take a job as a secretary and move on from there. Of course there are examples of secretaries doing just that and rising to senior editorial and managerial posts, and training schemes exist to net potential journalists in whatever department they may be found. But in the main, despite the size of the organization and the opportunities it offers, good secretaries quickly find themselves categorized as ... good secretaries, unable to make the leap into production without considerable difficulty. Similarly, think carefully before you go into research. In broadcasting it is usually a lower-graded occupational area, and not always considered part of the mainstream journalistic family. Far better, for example, to consider starting in a local radio newsroom, and aim to work your way across.

Writing in on spec

Between keeping a close and regular watch for advertisements for beginners, nothing is lost by writing in to employers on spec. An informal survey of eight senior journalists who recruit or strongly influence the recruitment of staff for television news programmes in the United Kingdom showed anything up to twenty unsolicited letters a week landing on their desks from potential young journalists, and all indicated their preparedness to offer interviews even if they had no immediate vacancies.

Unlike the United States, where applicants are encouraged to send a show-reel videotape, most of the British editors based their decision about whether to see one person rather than another on the strength of the CV and covering letter. 'Too few', according to one of the eight, 'incorporate show-reels, audio cassettes or cuttings.' Several stressed the need for the covering letter to be tailored to the employer – 'not an obvious photocopy of a standard application sent off to as many people as possible' – and to know something of the programme to which they were applying.

And if an interview is granted, what then? On average, those hard-pressed editors mentioned above were prepared to spare between ten and twenty minutes of their precious time talking to an on-spec candidate. With such a limited time available, it is essential to make a good first impression. Among the most consistent advice: 'be punctual', 'dress appropriately', 'have intelligent questions to ask', 'offer suggestions for stories', 'demonstrate awareness of the output' or, as one editor put it succinctly, 'Do homework'.

Finally, of the few basic qualities which would influence an editor's decision to put someone on a file for genuine future reference, three were most evident: originality of thought, ideas and tenacity (also defined as persistence, determination and commitment). Other important factors included experience and evidence of an ability to broad-

Getting into television news

1. Writing in on spec. Make sure you address your letter to the most appropriate person, usually the programme editor or news editor. Don't rely on names listed in reference books – they might have changed since publication. Check. Advertised posts will usually say to whom replies should be sent.
2. Covering letters. Rarely handwritten these days, although attractive handwriting will still impress many people. Use a PC if possible. (If you don't own one, borrow from someone else or ask the local library, your school or college for help.) One page, single-spacing, double-spacing between (short) paragraphs. Say clearly why you are writing and what you want. Get current and most relevant experience high up. Be sure to give your current address and contact numbers. Check spelling. Do not use a distracting font: stick to Times, Gill Sans or Arial.
3. CVs. Ideally: one page, typed clearly with single spacing, short paragraphs. Avoid fancy typefaces. Essential: tailor to the programme/organization in which you are expressing interest. A 'blanket' CV is not good enough, even if done professionally. Layout: personal details and contact numbers at the top, current and most appropriate experience very high up. 'Other activities: squash, theatre, socializing.' Not: 'Other activities: I am a brilliant kick-boxer player and have a wonderful sense of humour.' Unless, of course, you mean to be funny.
4. Photographs. Enclose a suitable portrait colour picture. Passport-size will do.
5. Show-reels and audio cassettes. Include one (only) if you want to 'perform', but it will convince no one of your talents unless pictures and sound are of reasonable quality. Record on a suitable format for replay, probably VHS for video and a mono cassette for audio. If in doubt, telephone the proposed recipient for advice. Rewind the tape to the desired starting point. Editors' desks are often awash with cassettes, so label yours clearly with your name and address. If you want it back enclose return packing and postage. Better still, keep the original and send copies only.
6. Cuttings. Consider sending a few by-lined samples, preferably duplicates or good photocopies. Never send precious cuttings books full of irreplaceable originals. They may go astray, and editors will rarely have time to read them anyway.
7. Filling in application forms. Type. Many application forms are constructed in a way to test your journalistic skills, so read carefully before completing. In this case sending in a CV by itself will probably result in rejection.
8. References. Enclose written comments from past employers if you think they will help. Before listing referees make sure you ask them.
9. Re-read. Double-check what you have written, paying special attention to spelling, punctuation and grammar. Typographical errors will suggest carelessness.

Figure 2.1

cast. The rest is down to intuition on the part of the editor, plus luck and timing. As one of the eight added: 'I found a presenter who happened to send in a video just as I was looking for someone.'

Direct-entry training opportunities

Apprenticeships, formal or otherwise, seem to have gone by the board, and the humble, once-classic old entry level jobs of copy-runner, tape-room messenger or gofer have to all intents vanished in the interests of economy, coincident with the arrival of linked computer systems and self-operated drinks machines. (There is a parallel in the development

A talent for television

Ability to communicate
Ability to distinguish fact from fiction
Awareness of what's required to get the job done
Basic experience, paid or unpaid
Demonstration of initiative
Evidence of commitment
Flair and imagination
Genuine interest in news and current affairs
Good all-round general knowledge
Enquiring mind
Knowledge of station/programme/subject area
Knowledge of how to nip around the Internet and how to use it to separate fact from opinion
Originality of thought
Persistence, persistence, persistence
Sense of humour
Solid grounding in basic journalism skills
Story ideas
'Team-playing' talent
Thorough awareness of potential of medium
Willingness to take on anything

Figure 2.2 Some of the main basic qualities expected of those trying to get into television journalism, as identified by eight senior journalists responsible for hiring or influencing the employment of entry-level editorial staff in the United Kingdom.

of non-linear editing, which is leading to the disappearance of junior assistant posts and, in turn, a lack of opportunity for new generations of picture editors to acquire the visual literacy previously learnt in part by watching skilled practitioners at work.)

In the United States, there is an established principle of 'internships', whereby graduates from college courses have the opportunity to test their newly acquired basic skills in professional surroundings as a forerunner to a potential career.

The British model is different. For a few there is the heady prospect of direct entry through full-time training contracts offered annually by the main employers. These schemes, which last anything up to two years, usually combine formal instruction with working attachments/placements to news programme areas. Guarantees of post-training jobs are rare, but sensible employers, conscious of the investment they have made in recruiting trainees in the first place, are inclined to look on them favourably when vacancies arise.

The schemes and the number of places on each are always going to fluctuate according to the likely employment prospects – there is, after all, no point in going through an expensive recruitment and training process only to find there are no jobs available to the participants at the end of it. Even the oldest, the BBC's prestigious News Trainee Scheme, which has produced a roll of some of the most distinguished editors, producers and reporters in the country over more than twenty years, had to be suspended at various periods in the 1990s when likely opportunities dried up.

Such direct training schemes, run not just by the BBC but also by ITN and several of the commercial TV companies, attract literally thousands of applicants, representing an enormous headache for those at the receiving end of the process. A deliberately difficult application form for the BBC News Trainee Scheme was constructed in an effort to deter the increasing numbers of entirely unsuitable candidates (cruelly dismissed in some quarters as 'the dog-handlers') and to ensure that only the potentially outstanding progressed to interview. Even so, sufficient numbers survive to keep the odds against success extremely high. That fact alone is enough to put off the faint-hearted, but remember someone has to make it: why shouldn't it be you?

Catering for special needs

For those already disadvantaged in some way, the barrier to success in broadcast journalism has been unfairly high.

Some ethnic minority groups have been under-represented in proportion to their percentage within the overall population of the United Kingdom. The mainstream broadcasting organizations were often seen as bastions of the white middle class, but efforts to change that perception are having success, with 'positive action' training projects aimed at the recruitment of black and Asian journalists in particular, and the emergence of role models in the shape of many international, national and regional presenters and reporters. In addition, the Multicultural Media Centre for the Millennium project started in 1999/2000 to put potential journalists from ethnic minorities through a series of fast-track training courses organized jointly by BBC News Training and ITN.

The attitude towards people with disabilities is also beginning to change. The sensitivity has extended to deleting the reference in some advertisements for journalists to the mandatory holding of a driver's licence, which implies full mobility. Confinement to a wheelchair should not presuppose an inability to think, write and speak, and employers who take the trouble to ensure that their buildings are properly designed to cope with such impedimenta are delighted to find they have tapped a rich seam of talent. Neither deafness nor blindness is necessarily a bar to a career in broadcast journalism. Several examples exist of journalists with severe hearing or sight difficulties who have made the grade in television as well as radio in the United Kingdom.

An additional fillip has come with the availability of bursaries provided intermittently by the Royal Association for Disability and Rehabilitation, enabling a limited number of people with disabilities of various kinds to attend recognized courses in journalism, including broadcasting, with a view to employment.

Trainee selection methods

The modern recruitment process embraces any number of techniques aimed at ensuring the right candidate is selected, and that applies as much if not more to trainees than to other potential employees. Lengthy psychometric testing, handwriting analysis, 'team-building' and group exercises, all designed to bring out character traits and deficiencies, may play their part, but a considerable weight – some say too much – continues to be put on the face-to-face interview. For the applicant this can be an intimidating experience, especially as the appointment process may call for large panels of interviewers sitting jury-like for several days in succession.

It is also worth being aware that every interview panel has its own group dynamic, with members as concerned with their own performance as with the candidate's. Some may be reluctant participants with other pressing responsibilities; some may be short-tempered for personal or professional reasons; others might be more interested in scoring points off their colleagues than anything else.

Whatever the composition of the inquisition, the need exists for candidates to be adequately prepared. That includes conscientious study of the programme area for which application is being made (obvious, but it is surprising how often those professing an addiction to a particular broadcast fail to make a point of watching the edition closest to the interview) and the display of a dress sense which matches the occasion.

A test of editorial awareness can also be expected, usually separate from the interview. General knowledge, written exercises, the selection and ordering of items in a simulated news bulletin, and some sort of journalistic legal/ethical puzzle are among the favoured ingredients.

Failure, if it comes, is not always final. Unsuccessful candidates should make a point of contacting the person responsible for convening the appointment panel. Feedback may well be encouraging enough to suggest it would be worth trying again.

Getting in from college

Although the qualities of youthful enthusiasm and the completion of a degree course are likely to be an advantage in competition for trainee posts they will, as we have seen, not necessarily turn the key to direct employment. At the same time, practical experience in the field is not easily attained, and there is a limit to the value of voluntary work. This perhaps helps account for the popularity of postgraduate courses in broadcast journalism. Several universities and colleges of further education now offer such opportunities, mostly of one academic year's duration, for a diploma which combines the study of law, public administration and similarly useful subjects with practical training in broadcast journalism using well-equipped studios and other facilities.

Courses also include short periods of working placements to radio and television stations, and these provide excellent 'shop window' opportunities for students. The professional teaching staff are often complemented by visiting current practitioners willing to lend a critical eye to proceedings. Some of these visitors come from local broadcasting establishments, thus creating mutually beneficial links with the colleges. For the broadcasters it offers a way of spotting the raw talent on their doorstep, while students have the obvious advantage of rubbing shoulders with the professionals.

This mix of the academic and vocational is endorsed by the industry at large, which monitors standards through the Broadcast Journalism Training Council. The BJTC, made up of representatives from the main employers, union and colleges, offers recognition to courses which seek it and are willing to adhere to comprehensive guidelines. Non-recognition does not imply low standards, but as the number of courses in broadcast journalism grows, the BJTC imprimatur is a useful guide for students looking for industry-approved training which they usually have to fund themselves. Many local authorities are reluctant to give student grants for broadcast journalism courses, but limited opportunities do exist for sponsorship.

It is significant that recognition has been given to or is sought by a widening range of colleges offering alternatives to the one-year diploma. A particular advantage of these courses is that most do not automatically exclude late developers, almost certain failures in direct applications for jobs, but who are not deterred by the prospect of starting new careers as virtual apprentices, working to programme editors several years younger. As a result, something in the order of 300 students a year come on the market through this route, and although the successful completion of a course does not guarantee a job, sufficient numbers win contracts or freelance opportunities to suggest that this is a better way than most of making a start.

National Vocational Qualifications

Training in broadcast journalism in the future may, however, be bound up in National Vocational Qualifications, the development of which gathered pace during the early 1990s. NVQs and their Scottish equivalent, Scottish Vocational Qualifications, have come about in an attempt to define acceptable standards of competence for workers within every industry in Britain. More than 160 industries finished the process well before the turn of the century, with broadcasting – joined, for the sake of this exercise, by its cousins in the film and video occupations – well to the fore, with the creation of a set of standards for each of its sixteen or seventeen occupational groupings.

NVQs are based on the ability to complete a series of tasks (competences) in the workplace. Defining those tasks and the methodology for assessment represented a mammoth commitment on the part of industry practitioners over more than two years. Establishing the framework while trying not to lose sight of the creativity, writing ability and news sense inherent in journalists was slow and frustrating for those inexperienced in functional analysis and much of the 'NVQ-speak' considered necessary had an Orwellian ring to it, but the journalism standards are now in place, with the first candidates seeking certification. There are several levels of NVQs, the lowest measuring the most basic of skills and the highest, currently level five, being suited to senior managers and professionals. NVQs for broadcast journalists have been pitched at level four, comparable with those working on British newspapers and magazines, the typical successful candidate being pictured as the producer (with a small 'p') of a news summary for regional television or of a bulletin in local radio.

Inevitably, perhaps, NVQs are being considered suitable for entry level only, although opportunities are there for existing practitioners to apply their prior knowledge, and it will take many years before they are universally accepted. NVQs may come to replace other existing vocational and professional qualifications.

Overall management of the NVQs for broadcasting, film and video is the responsibility of the industry training organization, named Skillset, which is also engaged in a continuing process to monitor the needs of employers, fund the training of freelances and generally help maintain and develop skill levels.

3

The electronic jigsaw puzzle

Among the inevitable first questions asked of all journalists by outsiders is 'How do you get your news?' Reporters do not spend their time simply patrolling the streets on the off chance of witnessing something interesting; neither is sole reliance on information from members of the general public a consistent way of filling anything other than a limited agenda. For those working in television the answer to the question is by no means a simple one, as so much depends on the financial and technical resources each news service is able to put into news-gathering, which is a complex and intricate operation in its own right. The small and poor, with little of their own to call on, may well have to be satisfied with second-hand material passed on by the sister radio services often run in parallel under the same roof, or the international television news agencies. The large, prestigious independents, with fat budgets, are able to bargain for exclusivity and/or make mutually beneficial deals to share coverage with like-minded non-competitors.

How television gets its news

Between the two extremes there lies a vast amount of common territory open to news organizations in general, whether they are engaged in putting the word out over the airwaves or on the printed page, for an audience of millions or a few hundred.

Much gathering and sifting of news is routine. Much of the news-gathering work happens in local radio stations, local news agencies and through Net-surfing. Regular enquiries are always made of police, fire and ambulance services, usually for news of immediate importance to a local region. The post produces its share of publicity handout material prepared by government departments, political parties, public relations firms, private companies, industrial and social organizations. To this rich harvest can be added 'house', trade and business magazines, official statistics, advance copies of speeches, invitations to exhibitions, trade fairs, inaugurations, openings, closings, the laying of foundation stones and other ceremonies of varying importance. Well-established fixtures – the parliamentary and political agenda, court sittings, state visits, sports events and anniversaries of all types – join the queue with scores of other public and semi-public events which are carefully weighed for their potential interest.

Those surviving the first hurdle are noted in diaries of future events for more serious consideration nearer the day. These so-called 'diary' stories or their immediate consequences (follow-ups) probably account for the majority of news stories which appear on television and in newspapers. The rivals to television news, whether they come in the shape of other newscasts, the Net, radio, magazines, daily or weekly papers, are scoured for titbits on which to build something bigger.

Staff, freelances and stringers

Journalists, working either for themselves or for other publications, offer suggestions (for which they expect to be paid) on a fairly regular basis. Freelances, or 'stringers' as they are called (and there are whole networks of them), are wooed by news editors against the day when a really big story breaks in their area. It is the stringers who provide much of the basic news. With good contacts among local police, politicians and business people, they are usually first on the scene of any important event in their community, and are swift to pass on the information. Local or specialist news agencies, concentrating on crime reporting, sport, finance and so on, also add their contributions, but it is the larger operators in this particular field who provide most of the bread-and-butter written information and still photographs on a regular basis to the broadcasting organizations and the press, with computer links to their customers speeding delivery on its way.

Subscribers obtain much of their domestic news from the British national agency, the Press Association (PA, founded 1868), which is owned by the chief provincial newspapers of Britain and the Irish Republic, and which supplies a complete service of general, legal, parliamentary, financial, commercial and sporting news. No serious rival to the PA has emerged since the Exchange Telegraph (Extel, 1872) stopped running a parallel general service to concentrate on financial and sporting topics. Reuters (London-based since 1851) is the main source of foreign news for the UK, with full-time staff based in more than 70 countries. Back in 1993 the agency made a powerful addition to its activities by taking a controlling interest in Visnews, then the world's largest television news agency.

The agencies themselves rely for their material on either full-time staff journalists or on the hundreds of stringers who owe first loyalty to the publications employing them. The end product of all their work can be seen every day in the hundreds of thousands of words which fill computer screens in every newsroom twenty-four hours a day.

The aim of all agencies is to provide their customers with fast (preferably the first) straightforward and accurate news of any event within their sphere of interest and to maintain their coverage for as long as the situation demands. Television, with its expensive and complicated news-gathering base, tends to weave agency material closely into its programmes, fashioning it into acceptable 'broadcast' style. Newspapers prefer to use the raw material as a reference point for reportage by their own editorial teams, although there are of course many occasions when the agency report appears alone or with limited alteration. These general newspaper reporters, traditionally under the wing of the news editor, are also deployed to cover the various diary items day by day, or may be detached for longer periods to work on projects or campaigns of special interest to their publications.

General reporters in television news are used mainly on the diary or follow-up assignments offering the most picture-worthy possibilities, and they are also engaged in spot news stories broken first by the agencies. But whatever their titles and however highly paid, generalists have always been fewer in number than their counterparts on individual national and big provincial newspapers, and are regarded in some quarters as a vanishing breed in the face of the appointment of subject correspondents.

These specialists, who concentrate their activities on particular areas of news, are regarded as experts after years of devoting themselves to a single subject and building up highly placed personal and trustworthy contacts, and in turn themselves become reliable sources of much that is important. The political editor, fresh from an off-the-record chat with a senior government minister; the economics correspondent, back from talks with people influential within financial, business or industrial circles – each is ideally placed to begin piecing together information which might well develop into a big news story, perhaps not today or tomorrow, but next week. This increasingly important group of specialist correspondents, as members of recognized groups or associations of journalists working in the same field for different news outlets, enjoy confidential lobby briefings from government departments, and are on the regular mailing lists of professional bodies sending out material of a technical or restricted nature.

The television news organizations are also able to rely fairly heavily on their own out-stations, which in turn employ specialist correspondents and staff and freelance journalists. Material originated locally can be pumped into the network news programmes live, or recorded onto videotape for playing at some convenient point later on. BBC News centrally is served by national production centres in Scotland, Wales and Northern Ireland, plus newsrooms in England. All are partly in business for the purpose of providing the national news with items deserving wider than purely regional coverage, although forays by teams from headquarters are by no means discouraged, and a recent development has been to base national reporters permanently in the most newsworthy regions.

It is also a two-way process, which allows regional broadcasting to tap into local stories which come out of London. A separate department at the BBC, complete with its own playout and swap-around facilities, is based inside the BBC News Centre in west London to help its regional outlets receive and exchange material. Facilities for regional services for both BBC and ITV companies are also at Millbank, a few hundred metres from the Palace of Westminster. So, for example, a Member of Parliament being interviewed on a subject of interest to people in one part of the country can be linked direct with any one of the nightly regional television news programmes or be questioned by a London-based political journalist on their behalf.

Yet another source is the BBC monitoring service at Caversham, about 30 miles west of London, which inputs into the BBC's news computer system information collected from a round-the-clock listening service to the radio and television broadcasts of foreign stations – often the quickest and most reliable way of obtaining international 'official' news.

Independent Television News has natural links with the newsrooms of companies making up the independent television network. It prides itself on an ability to act swiftly in the movement of people and equipment from London around Britain and beyond. ITN also provides news for Channel 4, Channel 5, and a substantial range of international news services for other broadcasters. ITN's studios, library, design and other facilities are also available to other production houses and broadcasters. Companies within the independent television network, who also put out nightly regional news programmes, exchange on-the-day raw material between themselves, and coordinate some coverage as a way of avoiding duplication of staff and equipment.

Sky News has been another force in British broadcast journalism, delivering a 24-hour service from its base at Osterley in west London. Sky makes considerable use of domestic and foreign news picture agencies, as well as having links with other broadcasters, but it also deploys its own staff reporters and camera crews. GMTV is also part of the domestic national UK news operation worth the job-hunter's consideration.

Foreign news sources

On the foreign side, some staff journalists are employed permanently away from base in any one of a number of important centres. These, by the nature of international affairs, are considered the most likely to provide a steady stream of news stories of interest at home: Washington, Moscow, Johannesburg, Brussels, Berlin, Jerusalem and Hong Kong, for example. Equally importantly, all these centres are themselves at the crossroads of international communications systems, but advances in technology have brought with them the flexibility for broadcasters to transmit their material from almost anywhere else in the world.

Maintaining a permanent base overseas anywhere is an extraordinarily expensive business, hence a recent tendency to develop bureaux and share some of the overheads with other broadcasters. Every correspondent needs transport, office space, ancillary help and somewhere to live. So the value of each foreign news bureau is kept under continuing close scrutiny for those who have to foot the bill for it, and changes are made to

Figure 3.1 Many journalists stationed abroad find it useful to have a place where they can socialize and swap gossip with other professionals. One typical watering hole for those in broadcasting and print is the Foreign Correspondents' Club in Hong Kong.

keep pace with the emergence of new areas of special interest. As a result, adjustments in the late 1990s have tended to produce more offices in Europe and southern Africa.

Like their counterparts at home, foreign specialists see that they understand the local language, meet the right people, mix with the representatives of other news organizations, read the local papers, watch television, listen to radio and get themselves accredited as the official representatives of their television stations at home. This, as a rule, will ensure a constant flow of information to be sifted for use as background material to items transmitted later on.

Although based for convenience in one place, the foreign correspondent might well have a huge territory, perhaps a whole continent, to cover. This means having to travel thousands of miles to reach stories breaking in remote areas. Time differentials frequently weigh heavily. The correspondent may have to work through the night to produce the goods for bosses for whom it is still daytime. He or she will probably wear two watches, one keeping local time, the other home time as a constant reminder of deadlines. With skill, experience and good fortune, the foreign correspondent will become an accepted part of the local scenery, sometimes as an honoured guest.

Elsewhere there may be hostility thinly disguised as toleration. The correspondent's home is bugged, the telephone tapped. Personal movements are monitored, contacts threatened or harassed. Eventually the correspondent commits what the government of the country regards as unacceptable professional behaviour, and is expelled. Reasonable comment or criticism in one country is regarded as sedition in another.

If that does not happen, after a few years in the same place, he or she is likely to be summoned home and reassigned by employers who fear the 'going native' syndrome effect on objectivity.

Where full-time television newspeople are not based, the foreign equivalent of the home stringer may be employed. Sometimes this is a local national serving any number of overseas outlets, but just as often the stringer is a foreign freelance accepting occasional commissions outside normal duties, or a staff correspondent of one publication which, in efforts to keep down the cost of maintaining a presence abroad, allows the supply of material to others, as long as priorities are maintained.

In addition to these permutations, there are also staff based abroad whose principal activity is to service radio but who – thanks to a policy which encourages the ability to operate across other disciplines – are able to turn their hands to television reporting and online writing when the need arises.

In some foreign countries the television news representative has easy access to locally based camera teams who are hired for a daily fee. In one or two particularly busy areas for news, crews are employed on a permanent or semi-permanent basis, and some of these provide highly sophisticated facilities which amount almost to studios in their own right. The material they generate is either shipped home by air for editing or transmission, although increasingly the nature or immediacy of the news dictates that it is handled locally, using hired staff, and then transmitted through the global communications system. This saves many hours and is being used more and more widely by those television organizations which see news as a highly perishable commodity and consider the outlay for hire of satellite time to be money well spent.

International news agencies

The other main users of the global system are the international television news agencies which supply the foreign material for hundreds of news programmes throughout the world. Their links with the main broadcasters give them immediate access to a staggering choice of first-class news material, and they also employ their own staff and freelance camera crews in important centres.

For the poorer news services, unable or unwilling to meet the expense of assigning their own staff to foreign stories, agency coverage is relatively cheap and rarely less than adequate in terms of coverage quality, whether broadcast in its entirety and supplemented by subtitles in the vernacular, or reduced in length and transmitted alongside a locally written commentary from the paperwork accompanying every tape.

The biggest agency is now Reuters Television, which began life in 1957 as the British Commonwealth International Newsfilm Agency (BCINA), undergoing a change of ownership to become Visnews a few years later. Reuters took a majority shareholding and changed the name in 1993. Part of its operation is to supply by satellite or video cassette a daily service of news and sport to more than 400 broadcasting stations. It was also involved in BrightStar, a permanent satellite link between Britain and the United States. In addition, Reuters has its own network of camera crews, many of whom have made an important contribution to the coverage of world events.

Reuters' main rival is Worldwide Television News (WTN), formerly UPITN, which was founded in 1967. At the beginning of 1994 WTN had fourteen principal bureaux around the world and camera crews in 87 cities, providing news and features by satellite and video cassette to about a thousand broadcasters in nearly one hundred countries. WTN also produces its own programmes, including *Roving Report*, a weekly half-hour current affairs magazine, series on the environment and health, as well as an annual review of the year. It also controls Starbird Satellite Services, previously shared with British Aerospace.

A new addition to the club came in November 1994 with the arrival of APTV, an offshoot of Associated Press, the long-established American agency. APTV's stated mission was to take advantage of its network of (at that time) 93 bureaux in 67

Figure 3.2 The Eurovision control centre of the European Broadcasting Union in Geneva, through which members make their news coverage available to others on a reciprocal basis every day. The material is free of copyright and can be used in news services without any restriction, the overall cost of the network being shared on an annual subscription basis. The centre moved from Brussels in 1993, bringing all EBU activities to one location. The EBU has set the pattern for similar cooperation between news organizations across the world. (Photo courtesy EBU: Piraud and Grivel; © Fabrice Piraud Geneva)

countries, adding dedicated television staff to provide a daily supply of world and regional news stories coordinated through its newsroom in London. CNN, although not an agency in the same sense, contracts to make its coverage available for transmission by other broadcasters. The main broadcasters are also sellers as well as buyers, striking reciprocal deals with the agencies and complementary news organizations. The BBC, ITN, Sky and the Japanese, American, Canadian and Australian networks, regularly form twin partnerships for the sharing of resources and news material. The relationships, however, are constantly changing, sometimes for editorial reasons and sometimes for solid business reasons. Much of the material offered by the agencies has the merit of being the first pictorial record of an important news story, to which the wealthier services may decide to dispatch their own teams later on. The agencies are also important sources for historical or background items for news programmes. Potted biographies, moving pictures and still photographs of events and places previously, or likely to be, in the news are rapidly made available when occasion demands, even though the material may consist of no more than a single portrait of an obscure politician or a few precious seconds of moving pictures.

Archives

Most television news departments maintain their own archives, largely built up from material they have already transmitted, supplemented by purchases from outside. These represent hundreds of thousands of separate news items, each carefully catalogued and indexed. Computer technology has been introduced to streamline the storage and retrieval process.

The once vast libraries of press cuttings and previously broadcast news scripts, complete with the smell of old paper and ink, were replaced in the late 1990s by software specifically designed to aid the breakdown of search words to as narrow a range as possible. Older newspaper cuttings (back beyond 1800) are available on microfilm. Old radio and TV scripts have not been destroyed but have been kept safe for the broadcasting museums of the future and are also sometimes needed as part of a graphic display for historical documentary features.

To this wide and fascinating mix of home-grown and expensively gathered news material can be added the occasional unexpected bonus – the tip-off from a member of the public, or home video which turns out to be a genuine exclusive for the organization lucky enough to get it. Modestly priced digital video cameras capable of excellent picture and sound quality are now so widely in use that some news programmes serving huge or remote areas actively encourage the creation of networks of enthusiastic amateurs on constant lookout for news items they can pass on for broadcast use.

Video news releases

Another sector is the 'video news release' (VNR), a video version of the printed press release. VNRs are made for companies and government agencies as a form of public relations, employing the expertise of television production companies. The product is professionally shot and edited and then distributed to television news services for use in their programmes. For hard-pressed editors, with rarely enough news camera crews available, VNRs represent a very useful source of pictures and information, but some journalists are also suspicious of what they see as a threat to editorial integrity. They fear some organizations will prefer to compile VNRs, over which they exercise complete control, rather than allow television news teams access to shoot their own material. Various pressure groups and campaigners will also offer video material to news organizations and, although they do so for non-commercial reasons, journalists still worry about how impartial the material will be.

4

Who does what in television news

If television news is a jigsaw puzzle, then most of the pieces represent highly skilled technicians with special contributions to make towards building up the final picture. For in modern industrial jargon television remains 'labour-intensive' which, despite the frequent trimming of staff numbers, means that a large percentage of what it costs to run such an organization is spent on the wages bill. Value for money is therefore essential.

Purely local broadcasting, frequently run on a tight budget, has always demanded nothing if not versatility from its people. In its mildest form this has been likely to give an executive senior managerial and administrative duties as well as editorial ones, an engineer responsibility for operating a television camera in the studio one day, a videotape recording machine in the newsroom the next, and a portable recorder on location the day after – or all three within hours. A journalist may have the task of combining news-gathering with news processing. In extremes this versatility requires reporters to edit some of their own pictures as part of a normal day's work and camera-operators to write commentaries for some of their own material. Journalists may also be expected to rotate between radio, television and text or online news services.

National news programmes, too, are not necessarily free of the need to be economical in their use of staff resources. One senior news editor in the Caribbean used to spend the first two-thirds of the day as a reporter and writer for the main evening news and the remaining third in the studio directing the cameras for the same newscast. It is not many years since a western European television news service operated without any permanent journalistic staff at all, apart from the chief editor. The writing for the main bulletins was undertaken by journalists who had already completed a full day's work elsewhere. Even among the more fortunate it was – and still is – not uncommon for television news to have to share such basic services as camera crews, picture editors, equipment and transport with sister departments within the same organization, even though obvious problems arise from the need to serve one master.

At the other end of the scale, the big national networks in television news usually have exclusive use of their own separate staff, studios and technical equipment. For example, the BBC operation is housed in its own wing of the News Centre, part of Television Centre in west London, but remains part of a public corporation responsible for a vast range of television and radio output. Independent Television News, with headquarters in Gray's Inn Road, is now owned mainly by a consortium of television companies to whom it supplies daily national news programmes for the ITV network,

(a)

(b)

Figure 4.1 Premises that have been purpose-built for news broadcasting organizations. (a) ITN's building in central London. (Photo courtesy of Wordsearch, Alan Williams); (b) The BBC News Centre is attached to the original Television Centre.

Channel 4 and Channel 5. Sky News is part of British Sky Broadcasting, delivering news 24 hours a day by satellite and cable to audiences in Britain and abroad.

Multimedia working and multi-skilling

Chiefly because of the demands on them to produce programmes at least three or four times a day or, in the case of the all-news channels, continuously, the 'big league' news organizations each need full-time staffs totalling several hundred. These are divided into smaller groups of specialists – camera-operators, reporters, picture editors, studio directors, graphics designers, newsroom journalists and so on, who have traditionally rarely strayed from a limited number of clearly defined duties. But changes are taking place under the influence of economic stringency, allied to increased competition. 'Multimedia working' has been heavily encouraged by employers anxious to maximize their human resources, and the drive towards newsrooms staffed by people expert in several disciplines has continued apace. Multimedia offices, with radio, text, online and television operating together, now exist in many BBC regions and within the substantial political and parliamentary unit near the Palace of Westminster.

Much the same approach is behind the move towards 'multi-skilling', the aim of which is to make use of what are called 'adjacent skills'. So, for example, reporters on location are encouraged to work microphones, picture editors learn to write, newsroom-based journalists are tutored in picture editing. The idea of an entirely multi-skilled, multimedia team is a very attractive one for employers seeking a flexible workforce, while for some workers the breaking down of the strict demarcation lines which sealed them into a range of useful yet often repetitive or mechanical jobs has been warmly welcomed.

Even some of the diehards recognize the logic of having one journalist covering a story for both radio and television, but point to the likelihood of being so busy meeting a string of deadlines that they have little time to exercise their news-gathering skills. Getting to the story, they complain, has almost become of secondary importance.

Other critics believe multimedia and multi-skilling are devices aimed principally at cutting costs and that in the long run the effect will be to damage hard-won craft standards. The trades unions involved in broadcasting are particularly suspicious of employers' motives and monitor them with an eye on the details of rotas, working hours and management behaviour. The unions were particularly worried about the Burn Out Factor. It was not so much a case of worrying about the news-gathering ability of a person who is required to shoot, write, and edit a story, but for how long that person can do it without his or her health suffering, alongside the problem of the quality of the final material. Many broadcasting managers started to realize towards the end of the 1990s that perhaps it was a bit much to ask at a time when the jury was out on pioneering late-1990s' digital technology. In general, journalists who had already mastered writing skills took much better to using the camera than they did to editing the vision and sound. Multi-skilling had started to settle in job 'families' (shoot–write or shoot–edit or edit–write or edit–direct) which provided a more stable approach to getting the news on air.

The organization of television news

With the advent of multi-skilling and bi-media working, the existence of any 'traditional' organizational structure within television news has come to take on even less meaning than it may ever have done in the past. Working practices vary widely across the world – sometimes within different parts of the same company – and the nuances of operation are such that a common language covering job titles and editorial/production methods has never satisfactorily taken root, glossaries notwithstanding.

That said, there is a similarity between many of the big organizations, which have functioned successfully for years by establishing a clear division of responsibility between those who gather the raw material for news (intake or input or news-gathering) and those who process and shape it for transmission (output), staffing the structure accordingly. This puts reporters and camera crews into the former category and newsroom-based journalists and production staff into the latter.

Operationally, both sides of the editorial machine are notionally under the control of the most senior output people on duty – usually the editors of the daily programmes. At one time it was fashionable to give responsibility for each day's news coverage, treatment and output to a single person known as 'editor for the day', the main benefit being seen as the continuity it provided, for each editor's spell on duty was 12 hours or more covering the transmission of several (shortish) news bulletins.

Editors for the day (more accurately, editors for part of the day) still exist in some all-news channels, and may control strands such as hourly summaries, but the current view generally acknowledges that it is virtually impossible for anyone to take charge of more than one main newscast a day, because programmes are longer, technically more difficult, and differ in style and content from others within the same stable. In any case, goes the argument, there would be scarcely enough time to give proper consideration to one newscast before the next one became due. So even though they may be sharing some staff and facilities, the editors of programmes within the same organization are usually working towards separate goals.

The news-gathering machine

Aside from operational duties on the day, the editor of a news programme within the intake–output system also has to shoulder a degree of responsibility for anticipating what will appear on the screen. Lengthy planning meetings are held daily, weekly and monthly, at which the meticulously compiled diaries are considered event by event under the guidance of the domestic and foreign news executives whose job it is to deal with the logistics of news coverage for the whole of the output.

The organization of television news

INTAKE/INPUT (News-gathering)	*OUTPUT (News processing)*
Home and foreign assignments editors	Programme editors
Reporters and correspondents	Newsroom editorial and clerical staff
Operations organizers (technical)	Programme presenters
Camera crews	Picture editors
Facilities engineers	Graphics designers
Dispatch riders	Video and stills archivists
Diary planners	Studio production and technical staff

Figure 4.2 Typical division of responsibilities in a large news organization, with one group of people concentrating on news-gathering, the other on news processing and production. Many roles are combined in smaller news organizations.

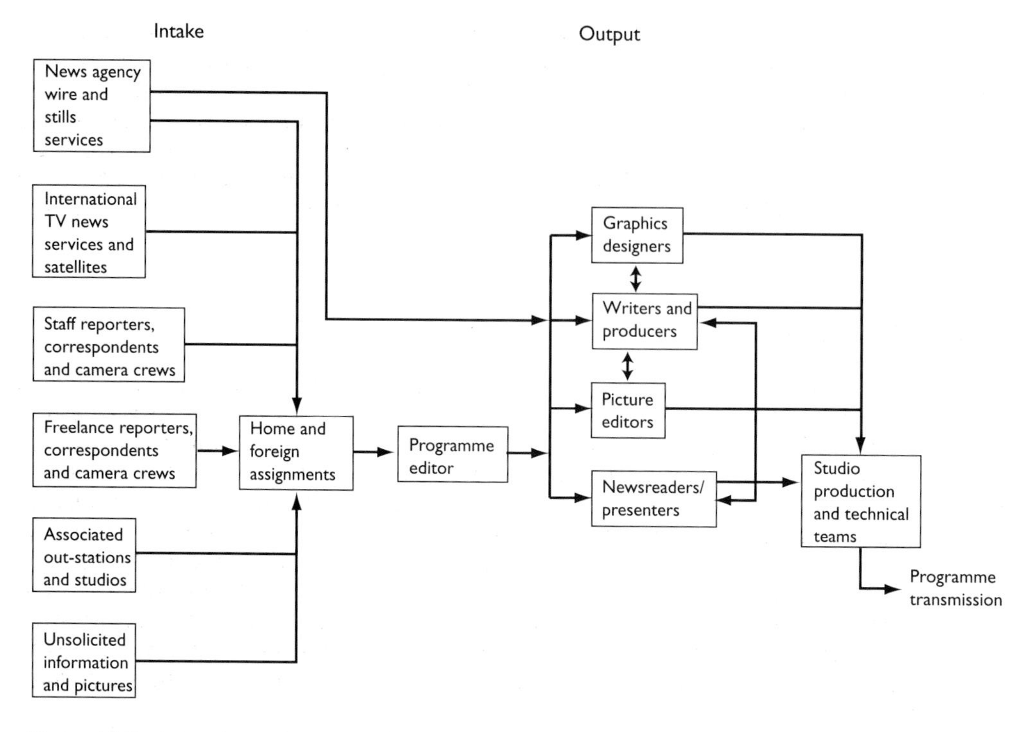

Figure 4.3 Progress of news items. Note the importance of the home/foreign assignments desks as the chief link between incoming sources and the output editorial and production teams. In some news services the role of writer/producer may be combined with picture editor.

At this stage each editor is really gazing into the crystal ball, trying to foretell what the programme on his or her next duty day will in part contain, even though it is clearly understood by all concerned that the most expensive, carefully-laid plans will be scrapped at the last moment should a really important story break unexpectedly. It is a hazard readily accepted by everyone, not least by those who may have spent many hours setting up interviews or obtaining permission to cover news items which may never be seen.

Such flexibility is a routine but essential part of the news-gathering process, which is relatively slow under even the most favourable conditions, although the speed of communication has improved considerably.

It has long been recognized that coverage for factual television programmes has to be organized well in advance to ensure that people and equipment are properly positioned as an event takes place. To that extent it is far simpler to call off coverage than it is to lay it on at the last moment.

Reporters and camera crews

At the sharp end of news-gathering are the journalists and camera teams, the troops ultimately responsible for providing much of the raw material. The reporters and correspondents 'get the story', conduct the interviews and do pieces-to-camera, while the technicalities are carried out by the camera crews. It is in this area that some of the most radical changes in television news have taken place. The introduction of very lightweight video cameras using cassettes which combine picture and sound has virtually ended the practice of two-man crewing. Camera-operators work alone, except for those increasingly rare occasions which necessitate separate sound recordists. A few lighting technicians/electricians also still exist in news, but their work is confined to those rare events which call for more artificial light than that carried routinely by the camera-operators.

Domestic assignments

The main responsibility for arranging news coverage eventually rests with the duty news editor or news organizer[1] staffing the home assignments desk as the mainstay of that part of the operation dealing with domestic subjects. With these journalists, through the programme editors and department heads, rest the moment-to-moment decisions of when to dispatch staff reporters and camera teams on assignments, or when to rely on regional or freelance effort to produce the goods. The news organizers often see themselves, somewhat cynically, as the 'can-carriers' for news departments, being criticized when things go wrong but rarely being praised for success. News organizers are meant to have the mental agility of chess grandmasters in moving pieces (in this case reporters and crews) into position before events occur, at the same time making sure that enough human resources remain available in reserve to deal with any important new events which may arise.

The work also demands a certain intuition about the workings of senior colleagues' minds. In briefing reporters, for example, they are expected to know instinctively how any one editor would wish an assignment to be carried out, down to the detail of questions to be asked in interviews. The role of the news organizer/editor is generally restricted to arranging on-the-day coverage, much of which is based on plans previously laid by other members of the department.

The duties of the planners include the submission of ideas for, and treatment of, the various items. But the routine calls chiefly for a well-developed news sense. This must be keen enough to isolate a tiny residue of screenworthy material from an overwhelming array of incoming information on subjects of potential interest. Much planning time is spent 'phone bashing', calling to arrange interviews, to verify whether what seems interesting on paper will actually stand up to the closer scrutiny of a camera, and to evaluate whether the various ingredients, as discussed, are likely to result in a clear and balanced report eventually being transmitted. Once the broad details have been agreed, each item, now formalized under a one- or two-word code name it will keep until it reaches the screen, is added to the internally circulated list of subjects for prospective coverage. At a still later stage, arrangements may have to be made to collect any useful documents or special passes needed on the day, so that the process of collecting the news may be carried out as smoothly as possible.

Planning for longer-term or particularly complicated assignments may well be conducted by special units created within the news-gathering department on a permanent or ad hoc basis. Big set-piece events such as summit meetings, elections, party political gatherings or any coverage destined to last several days, call for highly detailed organization in advance if the eventual reportage is to be comprehensive. It may be

1. These are operational titles. Broadcasting organizations have different titles for similar posts.

johnson mon 13 06 0719
=====================STRICTLY NOT FOR BROADCAST===================
NEWS DIARY MON JUN 13
AT 0800

EDITORS: JANET MATLIN (BREAKFAST)
PAT PALMER (LUNCH)
LES WINSTON (FIVE O'CLOCK)
CAROL DOBSON (EVENING)
DUTY NEWS EDITOR (HOME): BARRY LEVETON
DUTY FOREIGN EDITOR: CHRIS MOORE
DUTY LAWYER: STEVE KOSS

REPORTERS: HARRY HURST, DELIA WARD, LINDY ADAMS, AVRIL LEON
=====================================
OFF-DUTY: ROBERT GELLMAN, CYNTHIA YORKE
=====================================

HOME:

ELECTIONS	Post by-election Govt presser (1100) (Hurst) Opposition (1130) Prime Minister at Central Agents Assn meeting (1600) Political Ed upsum OB. Overnight scenes available
ELECTIONS/SCEPTICS	Sceptics gather for lunch at Old Boys Club (Poss 2-way via OB) (1245)
ECONOMY	Trade figs out (Econocorr watching)
DAMAGES	Record award expected by woman after car crash. High Court (1000) Poss interviews
OZONE	Scientists offer new finding showing hole has moved (Science corr to advise)
TOPFIELD	Oxfordshire farming pair facing strong Euro competition. Ward special for 5 o'clock.
HEALTH	Health Minister addresses one-day seminar. (Adv. copy of speech with News Desk. Note: midday embargo)
FOREIGN: EUROPE	Brussels reaction to election results
BRITISH BEEF	Agriculture ministers meet
REFUGEES	UN Sec-Gen speaks NY (Thompson)
CARS	Kansas auto firm promises revolutionary electric car. Poss pictures (Reuters?)
CRICKET	'Unknown' Aussie breaks Test record v India. Satellite + local commentary.

Figure 4.4 An example of daily home and foreign news prospects for a series of national programmes. This list of events for coverage is distilled by the planners from a wide range of possible stories for editors to consider.

PLANNING FOR PROGRAMMES

FIVE O'CLOCK REPORT: PROSPECTS FOR MONDAY JUNE 13

EDITOR: LES WINSTON
CHIEF WRITER: JANE BLACK
PRESENTERS: JOHN JONES, VAL WHITE
STUDIO DIRECTOR: PETER MANN

RUNNING ORDER MEETING: 1330
SCRIPT CONFERENCE: 1615

HEADLINES: John
HALFWAYS: Tony
ELECTIONS: Colin
ELECTION REACTIONS: Jane
ECONOMY: Jane
TOPFIELD & TOPFIELD EUROPE: Margaret
CRICKET: Jeremy
DAMAGES: Rob
KIDS: Ray (with Brian)
POSS NEWS ROUND-UP: Brian (with Ray)

Figure 4.5 In addition to the main prospects prepared and discussed at a daily editorial meeting, each programme team may compile its own as first confirmation of the main stories being covered, together with the allocation of the editorial team members to write and produce them. Knowing who is responsible for what makes collaboration easy for production and technical staff as progress is made towards transmission. The prospects will be made available on everyone's computer screens.

necessary to establish temporary headquarters away from base, committing substantial numbers of staff and technical resources to ensure that the main story and any side issues are properly covered.

The spin-off from all this effort may well be extended reports within the regular news programme, continuous coverage, or specials in their own right.

The same units may also be responsible for preparing background items and for keeping the profiles/obituaries of prominent personalities up to date – in fact for almost anything that helps programme editors avoid an unsatisfactory scramble to get something on the air.

Working in close harmony with the rest of the news-gathering department are the staff concerned with the technical side. An executive, variously named field operations organizer, assignments editor/manager, or camera unit manager works out rosters for the camera crews to go about their duties with or without reporters, probably keeping in touch by means of two-way radio telephone systems installed at base and in the camera vehicles. The executive's empire probably also includes the FACS (facilities) staff, engineers whose role is to supply the communications links between base and the camera crews on the road.

Other members of the intake team are likely to include clerks, whose duties involve booking studio time and facilities for material originating from regional and other outside sources, plus a small posse of motor cyclists who play an indispensable part in the collection of videotapes and still photographs.

Covering the world

Foreign news departments may have a small presence at base but this is the tip of a formidable iceberg made up of staff correspondents resident abroad, a world-wide network of stringers, and close ties with friendly broadcasting organizations able to conjure words and pictures from virtually anywhere in the world at very short notice.

The department is probably headed by a foreign news editor/foreign assignment editor, a senior journalist whose skill at balancing the books is becoming as big an asset as a sure editorial touch. Duties for foreign editors are likely to extend beyond company boundaries into a profusion of contacts with other broadcasters and the multinational 'clubs' established to provide regular, free-flowing exchanges of news material.

The routine administrative load is shared by deputy or assistant foreign editors, while operationally the better-off can afford to staff the foreign assignments desk with two or three foreign duty editors, working in rotation. These are the equivalent of the news organizers on the domestic side, providing a link between the news-gatherers in the field and programme editors at home. The dispatch of normally home-based staff on 'fire-brigade' assignments is a matter for negotiation between the more senior members of the department.

The last link in the foreign chain is the foreign traffic manager, or satellite operations organizer, who makes the detailed arrangements for the collection of material from abroad, and who needs to be in frequent direct contact with other broadcasting organizations, especially during the regular conferences between 'club' members. The residents of the foreign traffic desk are also renowned for their encyclopaedic knowledge of the procedures for organizing satellite communications at very short notice – a talent in demand when important news breaks at awkward times in some of the world's most unexpected places.

The three faces of output

Despite the phenomenal growth of news-gathering operations over recent years, television news output among the international players remains bigger and more complicated, encompassing as it does three services functioning within the loose categories of editorial, production and technical.

Among the craftspeople who work in these three areas lies the same sort of generally friendly rivalry which exists between a newspaper's journalists, advertising staff and printers. But, however much they may grumble about a perceived lack of appreciation, rates of pay and working conditions compared with the others, all are acutely aware of the fact that no single group's skills are by themselves sufficient to transform the material produced by intake into coherent television news programmes.

Much the same can be said for smaller-scale or differently structured programmes or services which, however organized, face similar processes in getting their product on to the screen.

The mechanics of editorial decision-making obviously differ from programme to programme. Logic suggests that policy is likely to be dictated by executives who are higher up the hierarchy and are expected to be in tune with the political, financial and commercial interests of the organization as a whole. This reflects the degree of autonomy delegated to the editor/producer, the senior member of the team responsible operationally for one newscast or other segment of airtime. Some editors are left entirely to their own devices (with the encouragement to refer contentious issues upwards) while bureaucracy confines others to work within more restricted limits. For these, mandatory attendance at regular editorial meetings chaired by executives not pressured by deadlines can be a frustrating and time-consuming diversion from other important duties.

Most programme editors are happiest away from their offices, preferring to base themselves in the newsroom where communications are most easily maintained.

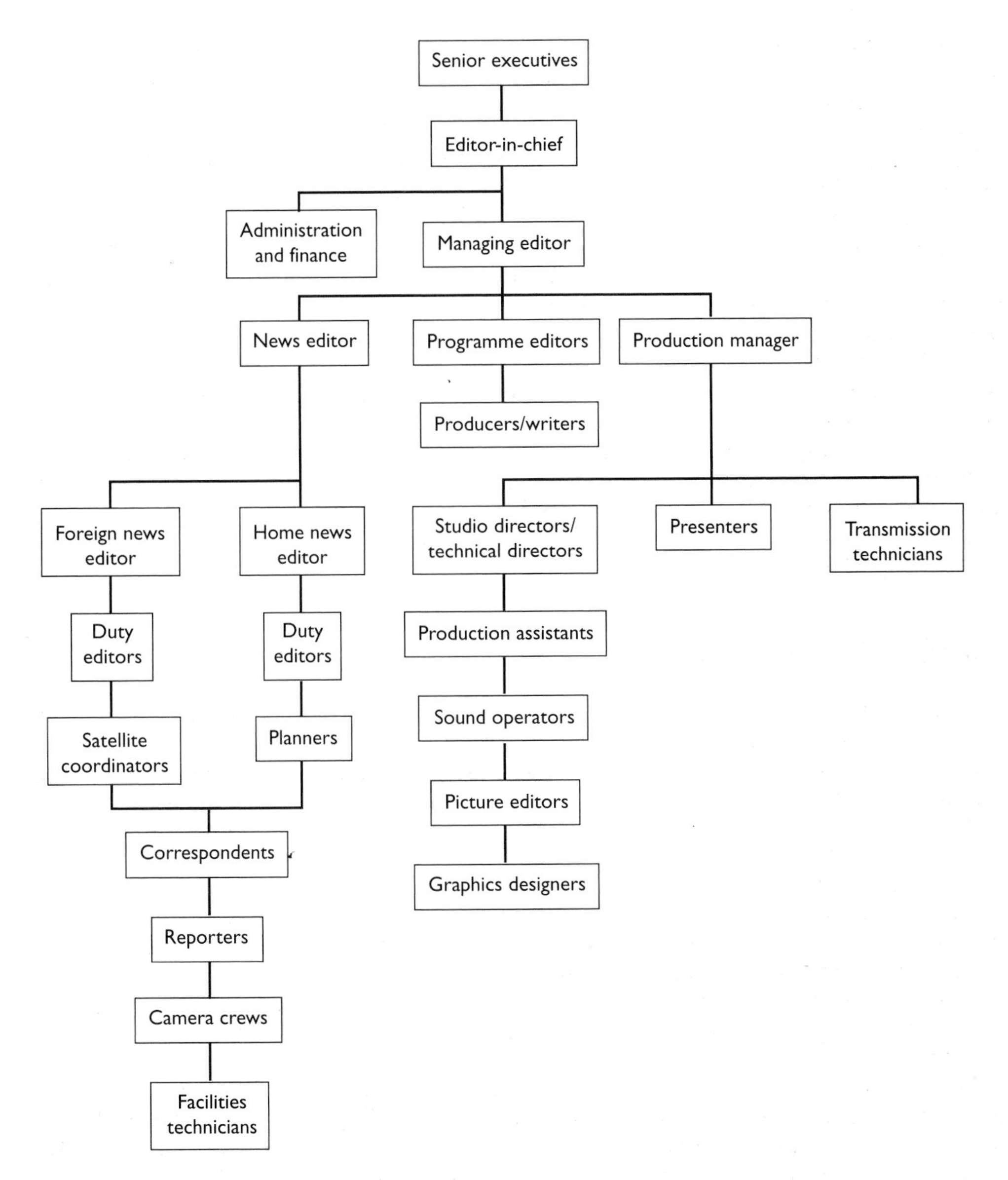

Figure 4.6 Television news organization chart. How a typical system might be structured. Hierarchies will differ, but in this example there is an Editor-in-chief responsible for all output to one or a number of senior executives. Administration, including staffing and finance, is carried out separately, while the day-to-day journalistic management is conducted under the overall guidance of a managing editor.

According to personal style, each editor exerts a different amount of influence to ensure that the programme follows its intended course. Computer terminals and television screens on their desk enable them to monitor the progress of coverage in the field and in the editing process, but much of the time is taken up with matters of detail affecting such things as content, construction, treatment, legal matters, taste and decency.

Marshalling the rest of the staff and assigning stories to individual writers may also be part of the editor's duties. Many are also expected to be the sole filter of raw material for broadcast as well as writing scripts of their own, assessing and checking the work of others and ensuring overall duration is within the allotted time span. A lot of attention is paid to the headlines at the top of the bulletin, not just the words but the images and sound that are needed to encourage the viewer to watch the rest.

Thinking time is at a premium for such busy people, and with so many calls upon an editor's time, and with so many loose ends to be tied during an often hectic few hours, responsibility for the detailed organization of a programme is likely to be led by a senior lieutenant. Titles vary, assistant editor/senior production journalist/senior broadcast journalist.

Although the chief's responsibilities may include some writing – perhaps the headline sequence, if there is one – the most important aspects of their work are likely to be as much managerial and administrative, with the accent on quality control and time-keeping.

Quality control begins with the briefing of the newsroom writing staff, perhaps six or eight people of different seniority, who are preparing the individual elements within the programme framework. It continues with the checking of every written item as completed so all the strands knit together in a way which maintains continuity while avoiding repetition, and language and programme style are kept on an even keel throughout. This may mean striking out or altering phrases, perhaps even rewriting entire items composed in haste by people working under intense pressure. Faster moving continuous news services do not always have that luxury and many items written by one journalist are never read by anyone else until they reach the presenter in the studio.

Also, as a journalist of great experience, the senior editor has further value as an editorial long-stop, preventing factual errors creeping through in a way which would ultimately damage the credibility of the entire programme.

Unlike continuous news services, such as Sky, CNN and BBC News 24, scheduled television news programmes (although agile when they need to handle a big breaking story) are rarely open-ended. They need to have a 'junction' with the next programme in the schedules. The senior journalist's talents are meant to include a facility for speedy and accurate mental arithmetic, essential for the other part of the role of time-keeper. Steering a whole programme towards its strict time allocation is no mean feat, especially as so much depends on intuition or sheer guesswork about the duration of segments which may not be completed until the programme is already on the air.

For these reasons the chief continually has to exhort those entrusted with individual items to restrict themselves to the space they have been given. A typical half-hour news programme might contain twelve or more separate elements varying in importance and length. An unexpected 10 or 15 seconds on each would play havoc with all previous calculations, resulting in wholesale cuts and alterations. These in turn would probably ruin any attempts to produce a rounded, well-balanced programme.

A further stage in the senior editor's time-keeping duties comes during transmission itself, when the appearance of late news may call for instant decisions on where to cut back. This means deleting material to ensure that, whatever changes have to be made for editorial reasons, the programme does not overrun its allocation of airtime, even by a few seconds.

Computer news processing systems can provide instant recalculations on programme duration, adding up not just the duration of pre-recorded reports, but also live reports that have just taken place and the duration of the links read by the presenter. Newsroom computer systems are now able to deal easily with all manner of arithmetical calculation, including late additions, subtractions, or wholesale changes to the order of transmission (see pp. 40-42).

Writers

The journalist newcomer to television news may expect to join the newsroom's pool of writing staff, usually the largest single group within the editorial side of output.

As the main link between news-gathering and the production and technical areas, writers exert considerable influence over what the viewer eventually sees on the air, substantial changes in their roles and responsibilities over the years keeping pace with variations in the style and content of the news programmes they serve.

These changes have been reflected in the different titles they have been given in the past, among them sub-editors, scriptwriters, news assistants, news writers and producers, the prefix 'senior' or 'chief' being added where ranking exists. Multimedia working has led to the introduction of the generic term broadcast journalist, denoting responsibility across radio, television and online or broadcast text services. On local stations especially, their work is almost sure to include intake duties, including the assigning of reporters and camera crews. Some may also appear on screen and in front of the microphone.

Such variations make it impossible to be precise about what writers do in every case, so the following description should be taken to apply in the most general terms. Within the limits of responsibility as defined by their job descriptions, writers/producers assemble the components which make up every programme item – selecting still photographs, graphics, artwork and videotape, writing commentaries and liaising closely with contributing reporters and correspondents. The most senior are often put in charge of small teams of other writers to compile larger programme segments from particularly complex news items made up of different elements.

What can also be said is that the complexities of modern television production are such that many writers find themselves with little time to devote to the actual business of writing, leaving it to their reporter and presenter colleagues.

Some writers go on to develop expertise in other areas of television technique and, as a result, are occasionally called upon to display their talents as field-producers, directors, or reporters on the screen.

In the busiest or more fully staffed newsrooms, other output duties may be assigned separately, whether or not they are recognised with formal titles. Where a programme uses electronically generated artwork a graphics producer may coordinate commissions and oversee progress until transmission; the responsibility of keeping track of the content and quality of moving pictures flooding in via satellite and from other sources may be delegated to a duty editor. While neither task (nor similar ones) may be considered strictly editorial, television journalists will argue strongly that, for example, the accuracy of spelling cannot be left entirely to a graphics designer, nor the news value of pictures decided solely by a picture editor.

Production output

Until the late 1970s it scarcely seemed possible to imagine the world's television news organizations exchanging their trusted old 16 mm sound cameras and fast-process colour film for lightweight electronic hardware and magnetic videotape. But they embraced the technological revolution with enthusiasm when it came and swept film aside in an astonishingly short time. At a slightly slower pace, digital cameras, editing and transmission systems started to replace electronic news-gathering and production during the late 1990s. Beyond 2000, digital is the norm: fast, light, cheap, mostly easy to use and in some cases disposable. The technology had a short infancy. Many senior news executives however will use the maxim Content is All – meaning that no matter what technology is used, it is that final material that comes out of the TV set or goes into your website that matters. Rubbish is always rubbish. The good stuff is always good. Don't blame the equipment.

Film editors, for example, became picture editors, adapting their skills to continue working closely with the writers from the newsroom, viewing and assembling the raw video material into coherent story lines within lengths dictated by their programme editors. Edited picture stories may run for a few seconds or several minutes, according to importance and, as with the editorial side, the more experienced are given the most complicated items to assemble. The junior staff handle those involving simple editing of minor stories (which may never get on the air) and material copied from the archives.

In the leading news services picture editors may expect to be allocated their work and generally supervised by senior colleagues with deadlines in mind. Other production staff are closely concerned with the operation of the studio control room, the central point through which all newscasts are routed.

The main occupants of the studio area (whether an integral part of the newsroom or a separate facility) are the newsreaders/presenters/anchors, the faces on whom the success or failure of any news service may be said to depend. Although reading other people's written work aloud for limited periods each day might not seem either particularly onerous or intellectually demanding, consistently high standards of news reading are not easily reached, and there are other pressures to offset the undoubted glamour of the job. In many ways the news presenter is well paid because he or she is supposed to be able to cope when things go wrong. They often do, and with very experienced presenters the viewer may never ever know it. Many of those with strong journalistic backgrounds are also closely identified with their programmes, to the extent that they now have a role in the decision-making as well as writing a large part of their own material.

In front of the presenter are between three and five cameras providing the link with the viewing audience. Many studios have robotic cameras operated from the control room which move to and from pre-programmed fixed points on the studio floor.

Although responsibility for the technical quality of the television signals being transmitted rests with a senior engineer, sometimes called the technical coordinator or transmission manager, the creative head of the control room on transmission is the studio director, who coordinates all the resources offered by the three areas of output. Helping fuse these together at the critical moment is a vision mixer to implement the director's selection of pictures from studio or other sources as defined by the script; and a sound operator/engineer to bring in the accompanying audio signals from microphones, recordings and additional soundtracks, etc. Slight errors or delays in reaction by any member of this team are instantly translated into noticeable flaws on the screen.

Several other creative groups may come within the category of production. Graphics designers/artists are engaged full-time on the provision, in accepted 'house' style, of all artwork used in television news. The software used by graphics is sophisticated, providing maps, charts, diagrams and the names of people appearing on the screen. The graphics computers can provide millions of colour combinations, more than the human eye could separate. Stills/picture assistants research and maintain a permanent, expanding library of photographic prints and slides, some of which are taken by staff photographers assigned to supplement material provided by freelances and the international agencies. Other librarians/archivists are responsible for keeping and cataloguing a selection of transmitted and untransmitted picture material. Newspaper and other print material are kept on computer files, although many older copied cuttings are still available.

Technical output

On a daily basis the technical staff are directly responsible for the maintenance and operation of both electronic equipment and complex computer systems. Their first aim is to ensure the highest possible technical standards, but at the same time to remain flexible in outlook, for compromise is often necessary where picture and sound material may be poor in quality but high in news value.

Engineering staff are also the mainstays of any number of units capable of transmitting news material directly back to base from outside locations. Links vehicles, fast response vehicles or satellite news-gathering systems are scaled-down versions of the multi-camera outside broadcast units used for the coverage of major events, and are integral to the process of news-gathering in an increasingly competitive environment.

Engineering tasks, though rarely sharing the limelight, are nevertheless central to the existence of any programme. Without them, and all the clerks, secretaries and other support staff working across the organization, the carefully constructed jigsaw puzzle of television news would fall apart.

Integrated newsroom systems

The argument against the intake–output system is that it is often the tail wagging the dog, concentrating too much power in the hands of the planners, who make most of the serious decisions about coverage, especially at times when money is tight. Editors often feel under pressure to use material which has been gathered at some expense, especially from abroad, even though their instinct tells them otherwise. Considerable financial responsibility goes with the coverage of any editorial 'patch'.

At the other end of the scale, modern integrated newsroom systems confine the decision-making process to smaller numbers. The editors in charge involve themselves as much in the detailed planning of coverage as in the eventual newscast for which they are responsible, and in some local US news stations still double as correspondents or anchors. Elsewhere, the editor may be that in name only, with authority stretching barely beyond the compilation of the running order.

The categories of non-broadcasting producer or writer may not exist, with every member of the editorial team considered a reporter, whether general or limited to foreign news, domestic, sport or other specialisms. Duties cover the whole spectrum of research, location reporting, picture-editing supervision and writing. In contrast to practices elsewhere (particularly in the United States, where scripts for national news programmes often undergo rigorous fact-checking and editorial approval processes before they are allowed on air), reporters are likely to have sole responsibility for content, treatment, script and duration. There is no consultation and supervision does not exist. They are allowed to be possessive about their material, and if they harbour any doubts about a particular aspect they will almost certainly discuss them with another reporter friend and not lose face by going to the editor.

Back in the 1980s, during the early days of their emergence from the Communist bloc, reporters from a particular country's television news service more or less chose their assignments for themselves. And although they were not strictly freelances, they were paid on the basis of what appeared on screen each night. In that currency two or three minor items of thirty seconds duration were worth more than a substantial report which took two or three days to compile. In this atmosphere, with the journalists keeping their own archive materials locked away for possible future use, the idea of professional teamwork and cooperation for the collective good was not exactly top of the agenda.

Newsroom layout

Although economics inevitably dictate the way a television newsroom is organized, the preferred layout for most continues to be open plan, a design thought to contribute more towards team spirit than a series of separate offices. Executives and some correspondents might be happier working away in complete or semi-privacy, but many journalists thrive on the buzz generated by a busy newsroom environment, and are prepared to put up with limited individual working space, noise and other distractions they would not tolerate in their personal lives.

Much careful thought goes into working arrangements, down to the exact positioning of workstations and the equipment to go on them. The development of multimedia duties, the relationship between the news-in and news-out parts, and the proximity of other areas of the operation, are seen as crucial to efficiency, especially when the newsroom also provides the background for news summaries or complete newscasts.

Computerized newsrooms

Computer technology began to replace the smell of ink and the clash of metal back in the 1980s. It started with word and news wire processing. Now computers are not only processing the scripts, but also aiding the editing of sound and vision and actually transmitting the entire programme.

Early systems concentrated on two main areas: the storage of incoming news agency wire stories for writers who were able to retrieve them with a couple of keystrokes, and then the use of the computer's word-processing facilities for scriptwriting, printing and collation. Running order composition and timing, reporter and camera crew assignments, duty rosters, news diaries and a host of other uses were added until the computer became an indispensable and integral part of news programme production.

Anyone familiar with the operation of almost any personal computer should be able to adapt easily to the demands of networked systems in which probably every occupant of a newsroom has access to a terminal.

Basic refinements include the ability to transform draft script pages into accepted programme format; a split screen, allowing wire copy to be retrieved on one side while

Figure 4.7 The computerized newsroom. Assignments, running orders, archives and many other functions are available to each and every journalist and member of the production team at the same time. Desktop editing of sound and vision also speeds the process of moving news from the news-gathering stage to its transmission. (Photo courtesy of and © Sky Television.)

a script is written on the other; automated calculation of script and programme duration, and access to archive material. Advances have also been made in split screens which incorporate moving pictures as well as text, while even greater scope exists for the establishment of links with the process of programme production and desktop editing.

For some time now it has been possible for scripts written in the newsroom to be read in the studio from cameras fitted with electronic prompting attachments. Using the later generation of newsroom computer systems, journalists are now able to compose simple graphics while sitting at their terminals and to insert instructions to operate stills stores and video inserts. The electronic news systems being used by most newsrooms now have menu displays, editing and powerful search facilities to enable the journalist to hunt a word or word combination, or to download all the scripts about a particular story.

One of the main benefits of the newsroom computer is its ability to cope with changes ranging from the smallest detail in a single script to the wholesale recasting of programmes. Amendments, additions and subtractions can be considered, entered and previewed at leisure, then executed in a twinkling. Running orders and completed scripts are available to everyone who is in sight of a terminal, at once. As long as the computer is programmed to tell the master printer the chosen final order, scripts for an entire newscast, of whatever duration, will be printed out in the correct sequence, and will be followed automatically by the electronic prompter.

Programme timing, too, becomes much simpler. The computer not only calculates the time taken up by each component of every script in a running order, it adds to the overall duration, remembering to adjust when new stories are added and others dropped, all the time taking account of a reader's individual reading speed.

Terminals sited in the studio control room remove the guesswork and arithmetical contortions from the process of transmission. As the broadcast progresses, the computer takes account of those items which have gone and those which remain to be transmitted, making minor adjustments to ensure that the programme ending on time is a simple matter for editorial staff.

For assignment desks, details of diary events and forthcoming coverage are entered by news planners and stored for retrieval by producers, reporters and camera crews when they come on duty. In the field, out-stations and overseas bureaux, laptops and desktops are available for reporters to write scripts and transmit them to base. The whole news process has been speeded up in a way never thought possible.

There may of course be a downside to all this. Getting out of a chair to indulge in conversation at the other end of a big newsroom may be considered unnecessary when it is so much easier to spend a few seconds in front of the electronic message-sender. Certainly there are many occasions when this will be a reasonable action to take, but care must also be taken to ensure that the habit of human contact between professional colleagues is not lost.

'Scripting', so much simpler using a word-processor, must not become a purely 'written' process, heralding a departure from the conversational style so crucial to broadcasting: writers who many years ago used to dictate their work aloud to typists became aware immediately if what they had composed 'sounded' wrong.

Another aspect to be considered is that health and safety managers must ensure that furniture and lighting are compatible with continual computer usage – some people complain of eyestrain, headaches and backache, as well as RSI (repetitive strain injury).

And the history of the future ...

History is a recycling of persistent truths. The journalist of 2001 and beyond is still faced with the same human-sized challenge as the journalist on a newspaper a century ago, long before even television news was invented. That is Garbage In, Garbage Out, or WYSIWYG (whizzywig). This means: *What You See Is What You Get!* What the journalist puts into the PC is what comes out in the viewers' faces, whether it is a bad script, bad editing, spelling errors on graphics or a keystroke hit in error which moves a story item to the wrong end of a bulletin. As mentioned earlier, Content is All. The computer won't write it for you and writing skills remain part of the human factor. Computers never make mistakes. They are machines. The most sophisticated computer system is not as complex as a single human brain and never as creative. Only people make mistakes, whether it is the people who write the computer programming, or the journalist who writes the news.

5

Writing for television news

A survey of almost one hundred radio and television editors in Britain by the Broadcast Journalism Training Council showed that the ability to write good English in the form of crisp, stylish scripts was the quality most sought after in the hopeful broadcast journalist. That may seem obvious, but there had been an impression in the late 1990s that good writing was not always appreciated. It turned out that this was not the case at all. Time and the proliferation of broadcasting services, it seems, have not changed the skills that matter. Good scripting and lots of ideas are what matter. The first thing to be said to the apprehensive newcomer about writing for television news is there are any number of broad guidelines but few hard and fast rules. This makes sound common sense in a medium where so much depends on instant reaction in the field or in the newsroom, up to and often including the time of programme transmission.

Television style

Attempts at some kind of standardization do, of course, take place from time to time, with varying degrees of success. Editors looking for continuity will occasionally assign senior journalists to produce lists of preferred spellings, titles and phrases to match the standards of ethical behaviour they expect from their staff.

All the big news organizations have their own style guides dealing with the way language should be used. The compiler of a recent style guide was a respected senior BBC journalist, but the guide took longer to produce than its slim 30 pages might appear. That was because almost everybody consulted had strong and often contrary views about certain elements: unwritten custom and practices being one thing, a book suggesting tablets of stone quite another. There was, however, no argument about the basic principles of television news writing:

- Be direct, simple and precise.
- Use short words.
- Use separate sentences rather than a maze of sub-clauses.
- Be brief. You write for the ear and eye.
- Prefer active to passive verbs ('he did something' rather than 'something was done').
- Use familiar phrases but avoid the tired ones.
- Prefer vivid language to the bland.

> It may seem obvious to say that broadcasting English is based on the spoken word, not written word. But try reading a few scripts under your breath and the good ones will stand out. If you're new to television, watch how many excellent writers do read to themselves. The good scripts sound as if the presenter is talking to the viewer, not just reading out loud ... Scripts are meant to be read by presenters from autocue ... complex numbers should be written out. Writing £17.4m is likely to make the presenter stumble. This should be written as seventeen-point-four million pounds.[1]

On my very first morning as a trainee reporter with a London suburban weekly newspaper, I was given a slim, yellow-covered volume to treat with the same reverence as I would the Bible. It told me, among other things, that the title Councillor had to be abbreviated to Cllr., Alderman to Ald., that it was not the High Road or even the High Rd., but the High-road, and so on.

In later years I worked in a newspaper office where the length of each paragraph had to be one sentence ... in another office, two. These sentences, it was made clear, must not begin with the definite or indefinite article, or with the word 'But', and it was a journalistic crime, punishable by office ridicule, to refer to the 18-year-old defendant in a court case when, all along, we knew him to be an 18 years old one.

All this may seem very trivial, and in many ways indeed it is. But ... but the fact remains, so far as the printed page is concerned, uniformity and consistency within the columns are more likely to please the reader than to repel. Haphazard changes of typeface or different spellings of the same word in the same or succeeding issues are guaranteed to irritate and annoy, and experts in design are much sought after to bring discipline and good order to newspaper pages, so that the reader's eye may be led smoothly from one article to another.

By its nature, television news cannot expect to do precisely the same. Complete control over every single word spoken by every single contributor – whether they are reporters, interviewees or participants in some other way – is unachievable in any practical sense, and while it is certainly desirable and possible to lead the viewer from event to event by the proper use of visual signposts, combined with careful phraseology, what occurs within the brief timescale of many a broadcast news item is open to each viewer's personal interpretation of what is being seen and heard.

It is this extra dimension which helps to place television in its unique position among methods of communication. A newspaper's verbatim report of an important political speech will give a clear record of what is said ... a newspaper reporter's word picture will give, at one remove, an interpretation of what is meant. A direct radio broadcast will enable what is said to be heard complete with repetitions, hesitations and 'bad' grammar. But only the television viewer, sitting in domestic comfort, is given the full information from which to make a personal assessment of the way things are said, together with the sidelong glance and nervous twitch which accompanies the confident-sounding delivery.

Unfortunately, it is extremely probable that the viewer will be shown a relatively small sample on which to base a judgement, as it is freely accepted by television newspeople that the 'whole' story, however important, can rarely if ever be told within the context of a routine 25- or 30-minute newscast sandwiched between the domestic comedy and the detective serial. Even if there were no pictures, and the presenter read continuously for an entire half-hour, it would not be possible to pack in more than about 5500 words – fewer than the front page of *The Times*.

1. *BBC News 24 Stylebook and Editorial Guide* (1999/2000). BBC.

However long they might like to linger over recounting events of the day, television news journalists are acutely conscious that, through no fault of their own, they have to be ultra-selective, both in the number of items put on the screen and the amount of time devoted to each. Critics outside and inside television news are convinced that these factors in themselves result in restrictions on the type of material which can be included, and believe direct comparison shows up a remarkable similarity in the content and treatment of news transmitted on the main programmes of the national broadcasting organizations in Britain.

Whether or not this is entirely true, there is little doubt that the time element does impose an important form of constraint on the newsroom-based journalist in particular. Yet whether it has only negative influence is arguable.

The need to condense forces continuing reassessment of the merits of individual items as they develop, ensures economy in the use of words, and discourages length for its own sake. Above all, it sharpens the newsman's or woman's traditional ability to recognize those facts which cry out for inclusion from those which do not.

The availability of all news 24-hour channels does not invalidate the argument. Continuous coverage of events as they unfold, however exciting live, unedited action may be, serves an entirely different purpose from tightly shaped reporting and explanation putting matters into perspective.

Application of a developed news-sense is only one half of the newsroom journalist's task. The other half, probably more important, is to convey the chosen facts in a way that every television viewer can readily understand. It does not mean pandering to the lowest common denominator of intelligence, but it does pose a problem which does not apply to the printed word. The newspaper reader fed with a regular diet of the tabloid *Sun* or *The Mirror*, for example, soon learns to expect every issue to be treated in the same bright and breezy style. The *Daily Mail* and *Daily Express* might be a bit more expansive. The regular subscriber to *The Daily Telegraph* or *The Guardian* has come to expect the treatment to be sober and more discursive. Television news wants to satisfy the readers of all of them.

Helping the writer do so is a powerful, double-edged weapon – the capacity to let the audience see and hear events for themselves. This advantage must not be squandered either by the use of technical wizardry for its own sake, so complicating otherwise uncomplicated issues, or by the presentation of written material in a way which appeals to only one part of the intellectual spectrum. Ed Murrow, one of the most outstanding of all broadcast journalists, recognized as much long ago when he urged CBS radio reporters to use language which would be understood by the truck driver yet not insult the intelligence of a professor.

It is true to say, of course, that the audience is becoming more fragmented both through the expansion of choice and a noticeable targeting by age and socio-political grouping, and experienced watchers of television news detect more 'downmarket' tendencies on some channels and a more serious, analytical approach on others. Clear differences of style and language, exaggerated by teenage, regional or ethnically-based cultural influences, are there for all to see.

Yet there should be no argument over the common ground which does exist. Today's sophisticated audience has become accustomed to hearing everyday words and phrases used in films, the theatre and on television. The news is no exception, and must be told in an authoritative, yet friendly and informal way which attracts and maintains interest.

Even though the audience may be made up of millions, the writer should be encouraged to think small, perhaps imagining people in groups of no more than two or three. Conversational language, preferably used in short, direct sentences, should be the aim.

Decide what it is you want to say: then say it. Experienced television journalists never forget their efforts will be totally wasted if the viewer does not immediately grasp what is being said, particularly when a moving illustration is competing with the spoken word. The admission may be painful to journalists, but the old cliché, 'one picture is worth a thousand words' has more than a ring of truth.

It is always possible for the newspaper reader to return to the printed sentence. If necessary he or she is able to pore over a dictionary. But words once uttered on television or radio are beyond recall. A viewer left wondering about the meaning of what has been said at the beginning of a sentence will probably be too distracted to comprehend what is being said at the end of it. That applies to every television news item without exception, and almost the greatest crime any journalist in the medium can commit is to leave part of the audience confused about what is meant. The onus is on the writer, always, to put across the spoken word in as clear, simple and direct a way as can be managed.

There is nothing to stop writers trying out their scripts on each other, because the essence of good writing for television remains fairly simple. In the course of everyday conversation, you would be most unlikely to say, for example, 'Chancellor of the Exchequer Jack Cash says he's given industry a multi-million pound boost with his cut in VAT'. On the other hand, you might envisage saying: 'The cut in VAT will help industry. The Chancellor, Jack Cash, says it's going to save millions.' At least it sounds more natural. And note the use of the word 'the'. Nobody normally refers to 'Prime Minister Tony Blair' or 'Film Director George Lucas' without the definite article. You wouldn't in conversation, so why should you do so as a writer for television news?

And there is usually no need to include the word 'today' in daily news programmes. Newspapers like to use it because it implies immediacy. In broadcasting, the assumption can be made that every story reported in today's news programme took place today – it's only worth mentioning if it did not.

It is, of course, far simpler to set down the principles of good writing than it is to carry them through, especially where some government publications, wordy official announcements or complicated economic or industrial subjects are concerned. Indeed official gobbledegook remains so prevalent that groups of language lovers exist to try to combat it. None of the founders of the Plain English Campaign thought they would still be coming across examples twenty years later.

> Despite all our efforts the problem never went away. As fast as we got rid of one form of gobbledegook another would raise its ugly head. And now new foes such as Eurospeak, management jargon and political correctness came out of the woodwork to confront us.[2]

Until the enemy is finally overcome, as a television newswriter you are left with a single, overriding test to apply. Do you understand what you are writing? If you do not, neither will the viewer.

The successful news script probably also has as much to do with mental preparation as it has with an ability to put words together in a clear way. The journalist working in television must be already attuned to the task ahead before anything is written. Watching yesterday's and this morning's programmes before setting off to work, listening to the radio on the way in and reading a selection of newspapers every day may at times be regarded as chores to be avoided, but the journalist who is not well-informed

2. The Plain English Campaign (1994). *Gobbledegook II – Utter Drivel.*

and up to date on a wide range of current subjects is unlikely to be genuinely authoritative when it comes to informing others. The sacrifice of 'pleasure' reading and viewing for 'duty' reading and viewing is an unavoidable necessity of professional life.

That every journalist should be keenly aware of what is going on in the world will appear to be stating the obvious. Regrettably an astonishing number are ill-informed about subjects they consider to be intellectually beneath them, and are proud of it.

There is also a proper routine to observe once the writer is given the day's assignments. Where applicable, there are the relevant newspaper cuttings, reference books and pamphlets to be consulted, coverage details to be discussed with correspondents from abroad or with reporters conducting interviews or constructing 'packages', and changes of emphasis to be watched on developing stories. Where pictures are concerned, a close check needs to be kept on progress from location to editing suite. In other words, unlike the newspaper sub-editor whose role is similar but not the same, the television newswriter does not simply sit still and wait for things to happen. When the moment comes to put words on paper or screen, the journalist should be in complete control of the shape and content for that part of the programme for which he or she is responsible.

Equally important is recognition that every writer's contribution, however vital, represents only one fraction of the newscast. There must be conscious awareness of the preceding and following items in the order of transmission so, where appropriate, the right phrases may be used to smooth the transition from one subject to the next.

Knowledge of what is in the rest of the programme ought to be automatic, but it is not. Editorial staff often admit they are so engrossed in their own particular duties they are completely unaware of what others around them are doing. It is a standing joke that the day will come when a writer handling down-bulletin pictures of a VIP opening some prestige project will be happily working away in an editing suite, blissfully unaware that, on his way back to the office, the same VIP had been run over by a passing steamroller.

Finally, although both the television and print journalist trade in words, what ultimately distinguishes one from the other may be seen as a matter of arithmetic. The newspaper or magazine sub-editor works in space – ems, ens, points and column centimetres on a computer screen or paper. The television writer works in time – minutes and seconds, and the formula that three words of English take one second for a professional to read aloud on the air provides the basis of all newswriting in television. This takes into account not only the slight variations in pace between readers, but also the different lengths of words used in normal, spoken language. It has survived the scepticism of successive generations of newcomers to television newswriting, and has proved itself both accurate and flexible enough to be adapted to other languages when calculated in syllables instead of whole words.

Broadcast language

One of the delights of the English language is its endless ability to absorb new words and meanings. For professional practitioners it represents a constant need to keep abreast of any current thinking and changes in usage likely to affect their writing. It does not mean accepting glaring grammatical or linguistic errors just because they are in common use, but, for example, to use the word 'gay' in its original sense would be flying in the face of reason. On the other hand, my own view is that unnecessary Americanisms have also crept in. Thankfully, while 'sidewalk' and 'diaper' have yet to be established in place of 'pavement' and 'nappy' the term 'meet with' is heard regularly, as are many other American words or expressions.

The transatlantic influence on UK journalism may also be responsible for some of the gibberish which has crept into the language of business and management. Regrettably the broadcast industry, whose primary objective ought to be good communication throughout, is not immune. Among my personal prejudices are 'downsizing' (sacking people), 'movement sheet' (diary) and 'human resource departments' (personnel).

A more serious argument must be made for the exclusion of sexist, racism and ageist language.

Racism is unacceptable, full stop. The multicultural society is now a fact of life, and professionals addressing a general viewing audience would be foolish not to remember to reflect as much in any writing. Careless or unthinking references to colour, race or religion can be as hurtful as deliberate ones. Only if someone's racial origin is strictly relevant to the story should it be included.

While age and experience are rightly revered in some cultures, the phenomenal dash for youth, especially in the Western media, seems to have brought with it a certain disregard for anyone over 40. (A nameless very senior manager in one of the world's leading broadcasting organizations is said to have described anyone over fifty as being 'brain dead'.)

It is a mistake. The age group between 40 and 55 accounts for around 11 million members of the UK adult viewing audience of just under 53 million[3] and in total has a huge amount of disposable income – and votes – at its command. And while 'middle-aged' and 'elderly' might be perfectly acceptable terms, they are often applied inaccurately. Of course it is possible to go too far. It has been said that in Canada, for example, a 'senior citizen' (pensioner) might be described as a 'Mature Canadian' – a term which might just as easily be taken to refer to a ripe Cheddar variety of cheese.

The battle for sexual equality has been fought and largely won, although many women would say it has not gone far enough. It is no hardship for any writer to acknowledge the importance of both sexes by referring to 'police officers' instead of 'policemen', 'fire-fighters' instead of 'firemen', 'pilots' instead of 'airmen', and so on. Where it becomes more difficult are those occasions when the use of neutral gender is likely to raise a smirk among a large enough proportion of the audience to make it matter. 'Spokesperson' and 'chairperson' sound as wrong to some ears as they do right to others. It has, of course, to be a decision for each news service if not each writer, but a middle way might be simply to identify the individual 'spokesman' and 'spokeswoman', 'chairman' and 'chairwoman' when they can clearly be seen as such.

For the writer, the uncertainty of political correctness, or what is 'right' to say to the viewer represents a minefield. Those who argue sincerely for proper acknowledgement of minorities, the under-represented and the disadvantaged have a point. The description of a wheelchair user as 'disabled' or 'handicapped' assumes mental as well as physical impairment and it requires only a modest change of emphasis to improve accuracy and avoid offence. There are many other examples.

The trick is to acknowledge change without having the average viewer hooting at the screen in derision. It is hard to imagine 'vertically challenged' (short), 'client of the correctional system' (prisoner) or 'terminally inconvenienced' (dead) and other terms[4] becoming part of everyday language. Equally, the string of words and phrases banned by the *Los Angeles Times* newspaper in its *Guidelines on Ethnic, Racial and Other Identification*, issued to staff several years ago, included 'inner city', 'Hispanic', 'ghetto' and 'male nurse', none of which would seem likely to strike British ears as immediately offensive.

3. The total adult viewing audience is defined as those aged four and over. Broadcasting Audience Research Board figure for 1994.
4. Henry Beard and Christopher Cerf (1993). *The Official Politically Correct Dictionary and Handbook*, Grafton.

The American influence

English-English	*American-English*
Caravan	Trailer
Caretaker/nightwatchman	Janitor
Cargo	Freight
Chemist	Drugstore
Chips	French fries
Film	Movie
Goods train	Freight train
Hire	Rent
Holiday	Vacation
In hospital	Hospitalized
Lift	Elevator
Lorry	Truck
Postman	Mailman
Meet	Meet with
Mortuary	Morgue
Piller box	Mailbox
Ring/phone	Call
Rubbish	Garbage/Trash
Solicitor	Lawyer
Stones	Rocks
Timetable	Schedule
Transport	Transportation
Windscreen	Windshield

Figure 5.1 English is a living language, capable of absorbing new words and meanings, but large parts of the viewing audience may consider the use of Americanisms unnecessary. The above represents a sample of those heard on British television news programmes.

Reader on camera

The simplest way to present the news on television is for the writer's words to be read direct to the viewer through the electronic camera and the microphone in the studio. In the terminology of television, the reader becomes on camera (on cam) or in vision, which makes the written item a to-camera story or vision story.

All television news programmes contain varying numbers of vision stories. Sometimes they are complete in themselves; more often they are used as a base from which the presenter launches some visual material, hence the frequent use of the terms vision/on camera intro, lead in or link. Theoretically, although an in-vision item may be of any duration (it is in any case impossible to generalize about 'ideal' lengths) editors of news programmes have a tendency to keep them to within reasonable limits for fear the programme presentation as a whole may seem to lack pace and variety. It is also felt that long vision stories, those going much beyond a minute (180 words) or so do not make the fullest use of television's possibilities. Conversely, there is believed to be little point in producing a vision story shorter than two sentences, as anything less seems unlikely to register with the viewer.

Superficially, there may appear to be very little difference between a vision story, newspaper article, or news agency raw material. In fact, there are essential variations. The opening paragraph of any newspaper item will make a point of establishing four main facts – who, what, where and when, as in the typical example.

Mean what you say

Euphemism	*Meaning*
Air strikes/air support	Bombing
Collateral damage	Civilian casualties
Deprived areas/inner cities	Slums
Developing countries	Poor countries
Downsize/streamline	Give people the sack
Early/premature retirement	Redundancy
Efficiency squeeze	Sackings
Efficiency savings	Sackings
Flexible workforce	Short contracts
Friendly fire	Killed by your own side
Human resources	People
Interpersonal skills	Getting on with others
Job losses	Redundancies
Legitimate targets	Police, soldiers
Lower income bracket	Poor
Majority countries	See 'Third World'
Made aware of their responsibilities	Threatened with the sack
Measured response	Retaliation
Movement sheet	Diary
Neutralize	Kill
Reconstruction	Sackings
Reordering of priorities	Budget cuts
Reporting guidelines	Censorship
Senior citizen	Pensioner
Soft targets	Defenceless civilians
Substandard housing	Slum
Take out	Kill
Third World	Poor countries

Figure 5.2 Euphemisms. Some jargon words or phrases employed by politicians and others to mask unpleasant or unpopular actions, or to avoid offence, have crept into the language: writers for television should be sure they are not doing the same thing.

> Luton, Bedfordshire, Thursday
>
> Two masked men armed with shotguns forced their way into the High-street branch of Pitkin's Bank here this afternoon and held staff and customers hostage for nearly an hour while two other members of the gang stripped the vault of an estimated half a million pounds in cash and jewellery from safety-deposit boxes.

An attempt to follow precisely the same pattern on television, with so many facts packed into a very short space of time, would almost inevitably lead to confusion in the viewer's mind. Arranging the same facts into a different order, the television newswriter aims to explain the incident in much the same easily-understood language which would be used to a group of friends.

> There's been a big bank raid in Luton, about thirty miles from London. An armed gang held customers and staff hostage for nearly an hour before making off with cash and jewellery worth about half a million pounds. The bank, a branch of Pitkin's ...

Using a similar technique, such complicated subjects as trade figures or retail prices indices need hold no terrors for the writer, even if both sets of statistics were to arrive for publication at the same time:

> The economy's continuing to show signs of recovery. New sets of official figures show that for the sixth month running Britain sold more abroad than ever before, and prices in the shops have dropped again.

Once that basic message has been put across, the details can be added with charts prepared by the graphics department. But it is accepted that opening sentences do represent one of the most difficult areas for writers seeking a compromise between impact and full comprehension. There is, for example, nothing much wrong with this sentence which might be heard on a typical regional news programme:

> Railway fares in the South-east are going up by an average of ten per cent in the autumn.

Yet if heard with anything less than full attention at least one of the four facts may be missed. The alternative leaves little margin for error. First, the viewer is hooked:

> Rail fares in the south-east are going up again.

then firmly landed:

> The increases, averaging ten pence in the pound, take effect in the autumn.

Of course there are occasions when this approach would be considered far too soft and tentative. The television newswriter must then talk in bold headlines:

> Five hundred people have been killed in the world's worst air crash.
> The Government has been defeated in the Commons.
> And Manchester United turn pain into pleasure in ninety seconds in Munich.

Notice how the last one can be a tease, forcing viewers to stay tuned to see what that means.

The transition from these to less momentous events is sometimes best achieved by the use of a form of words most easily described as side headings, to signify change of pace and subject:

> Next, the economy ...
> At home ...
> Abroad now ...
> In Munich ...

In some quarters these and similar phrases are beginning to be considered clichés, but, used sparingly, they remain good examples of the kind of language which in principle can be used to lead the audience from one item to the next.

There also comes a moment in an event which has been reported continuously over a period when it is desirable to resort to easily understood shorthand by way of an opening phrase. In that way, viewers came to recognize that, for example, 'The Northern

Ireland peace talks...' was referring to the drawn-out negotiations between the political parties in Northern Ireland during 1998–9. The invasion of Kuwait by Iraq in the early 1990s, which led to fighting between the United Nations and the forces of Saddam Hussein, was termed 'The Gulf War' almost as soon as the first missiles were fired. In each case these few words were sufficient to make members of the audience sharpen their senses in preparation for the latest news of events about which they already had some knowledge.

In the context of short news programmes, this shorthand technique is most effective. But there are clear dangers: first in the assumption it makes of the viewer's understanding of what has already happened; second, in the case of complicated issues, the background may easily be forgotten. After a while, for example, the inquiry which followed the leaking of a secret report into the treatment of inmates at the Naughty Boys and Girls Institute of Correction is handily telescoped into 'The Naughty Boys and Girls Inquiry'.

A week later, with the inquiry deep into evidence from parents and staff, complaints from social workers that previous warnings of bad treatment had gone unheeded, and questions to the government in Parliament, the viewer may be forgiven for losing sight of the story's origins. So in these cases it is necessary to go back over the entire ground, however much some editors might consider it a waste of precious airtime.

But, whether it concerns an old event or a new one, the writer's aim must always be to ensure that an opening sentence of a vision story hits the target first time. The viewer must be properly alerted to matters of interest and importance by the skilled use of words which, in their effect, have the reader bawling from the screen: 'Hey, you! Watch this!'

Adding illustration

The first stage in making the straightforward vision story more interesting for the writer to construct – and the point at which television starts to exploit its inherent advantage – comes with the addition of non-moving illustration to take the place of the reader's image on the screen while the voice continues to add information.

Almost all illustrative material is stored on computer. Graphics has become the term most widely applied to this material, encompassing as it does both artwork and anything originally photographic. Years ago the stock 'personality' pictures to be found in television news libraries resembled the dull, full-face mug shots usually seen in police records or staring from the pages of passports. Then the spread of colour television in the late 1960s and early 1970s gave an opportunity for a complete rethink. As a result those ugly passport snaps gave way to bright, natural, frequently unposed pictures taken by photographers mindful of the television screen format, which was wider than it was high. The arrival of widescreen television sets and widescreen transmissions in 1999/2000 forced another rethink about stills, which needed to be trimmed into frames by graphic designers.

The bulk of these stills start life as 35 mm colour transparencies, although some colour prints, including instant pictures for speed, are also used. Government departments, embassies, specialist and trade libraries also provide a mixture of shapes, sizes and qualities. The international picture agencies, while remaining easily the most important source of black and white prints, particularly from abroad, are also able to transmit colour pictures by wire through a tri-colour separation process. An original colour picture is re-photographed successively through red, green and blue filters, and the

Figure 5.3 Left: old-style 'mug-shot' could have been taken from a passport; right: modern-style photograph, natural and unposed.

results sent over the wire. The original scene is reconstituted in full colour by reverse application of the same process in the darkroom or by computer.

Rather less complicated and cheaper methods of acquiring stills include the 'freezing' of video frames, which can then be edited as part of other sequences or manipulated by computer.

The digital frame-store

Old photographic methods have been made more or less obsolete. Nothing did more to improve 'the look' of television news programmes in the late 1990s than these incredible pieces of graphics design equipment which have introduced almost limitless variety and versatility to the humble still picture.

Nearly all systems of this type depend on the concept of the frame-store, a bank of computer memory in which an entire (still) television picture is stored. Each picture element or pixel – and there will be more than 400 000 in a single frame – is converted into a digital representation of its brightness and its colour. The information is then stored in the memory – the digital frame-store. From there the picture can be replayed, recorded or modified, with virtually no detectable loss of quality.

A typical system for recording and transmitting stills would have two frame-stores and a magnetic disk memory just like the Winchester disk machines used with large computers.

In its simplest form, pictures are taken into a frame-store from a video source – a camera, videotape player or digital source for example – and then transferred to the disk memory under an identifying number, a single disk holding perhaps between 250 and 300 frames. Any still frame recorded on the disk may be called up by its number, transferred into the frame-store and offered as a picture source to the vision mixer in the studio control room, or direct to a videotape machine, or even another frame-store.

The refinement offered by two frame-stores allows one to provide the picture for transmission while the other is being loaded with the next selection from the disk, a process of little more than a second. If two monitors are connected to the output of the frame-stores, one will show the 'on-air' picture, the second a preview of the next available still.

A video switch between the two frame-stores then makes it possible to cut directly from one to the other without the need to come back to 'vision'. The next logical step is to arrange for the 'off-air' frame-store to be loaded automatically as soon as it is free with the next picture from the disk. In this way a series of presses of a single button will transmit a sequence. A further facility may be provided to allow the changeover from one frame-store to the other to be an electronic mix or wipe as an alternative to the usual cut.

To take full advantage of these options it is necessary to have some way of putting the pictures into the order required for transmission. One way to achieve this is by what resembles the 'Polyphoto' technique, a sort of mosaic of sixteen or more pictures from the disk displayed simultaneously at reduced size, together with their catalogue numbers. Using a keyboard and cursor on the screen, the operator may then move, insert or delete pictures to obtain the sequence he or she requires.

It is perhaps interesting to note that in operations of this kind the pictures are not actually transferred from one part of the disk to another. Instead the disk's directory system merely reorganizes the references to where they are to be found for retrieval in the desired order.

Stacking and cropping

In yet another refinement, dozens of sequences of stills may be recorded by the machine in groups or 'stacks' of perhaps as many as 80 pictures each. A stack may itself be one of the items (instead of a picture) within another stack. When the machine encounters one of these 'nested' stacks it reproduces the pictures in order, and then returns each to its place in the original stack. This facility is also provided to modify previously prepared stacks at the time of transmission, so that items may be inserted or skipped to allow for late changes in the newscast.

The frame-store principle also makes it possible to alter the framing and size of a picture. To 'crop' a picture, a rectangular outline is superimposed on the screen. This rectangle may be altered in size, proportion and position to frame any part of the image, and that information is stored together with the picture to which it applies. Thereafter, when called upon to do so, the machine will offer that image cropped in the memorized format, if necessary reduced to fill the space allotted to it.

A similar procedure allows the creation of composite pictures. A background may be formed by using an existing picture or one of a number of colours provided, and the cropped image then 'inserted' into it. If a composite picture is first stored and then used as the background for a subsequent composite and that process repeated, the resulting stored frames may be made into a stack.

Storage and retrieval

The number of pictures it is possible to store will usually be limited only by the number of disks which can be afforded or accommodated, and that in turn will influence the size and complexity of a management system to keep track of the contents. Dates and titles can be added and the pictures retrieved by number, name or category, but in the end the effectiveness of a library system such as this, particularly as it grows, will depend a great deal on methodical house-keeping – clear titling and cataloguing backed up by a ruthless weeding out of unwanted material.

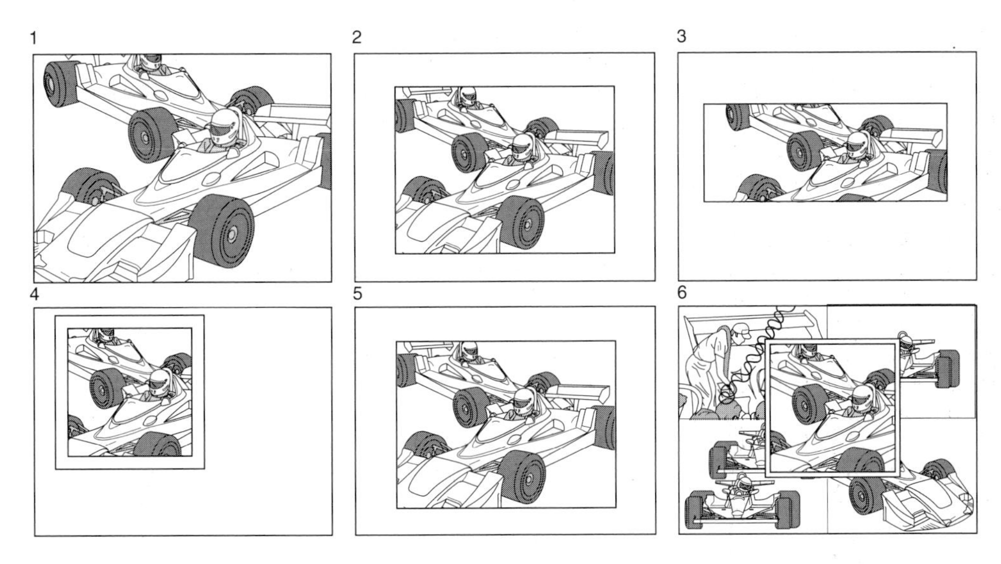

Figure 5.4 Pictures can be taken from any video source, stored, cropped, shaped or moved to fit into any background or montage. 1, full-size picture; 2, picture size; 3, crop; 4, picture position and border; 5, matte; 6, montage.

Making the choice

In many respects it is far easier to choose a still from a strictly limited range, for if a collection carries only one picture of a person or subject there is a straight choice between using it or not. The headaches begin with the vast libraries. Between them BBC Picture Archives, CNN, Reuters, Sky News and ITN carry millions of pictures, both as 35 mm colour slides and as computer-stored images. Such luxury demands the newswriter's intimate knowledge of each subject as it arises, for no one else is in a position to decide which of the available selection most aptly matches the mood of the story. Judgements may be as basic as ensuring that an item about increased taxes does not show a broadly grinning finance minister who has just announced them to Parliament in the face of strong opposition. That decisions as simple as this are needed only goes to prove what traps exist for the unwary.

For while straightforward mistakes in identity could in some circumstances have legal implications (and have, on occasion), even where the matter of identity is not disputed a writer's preference for one image over another could lead to accusations of distortion. Subtle changes in the appearance of leading personalities – hairstyle and colour, age, weight, and so on – all need to be noticed, although it is not simply a matter of choosing the most recent photograph, and a constant weeding-out process is necessary.

Writing to still pictures or insets

A straightforward 'personality' still picture should be on the screen for a minimum of five seconds (15 words). Anything significantly less is likely to be subliminal for the viewer, for whom the image will seem to vanish from the screen before it has time to register.

The maximum time depends very much on the subject. A fairly 'busy' action shot of casualties being carried by stretcher away from a train crash needs longer to register than the library portrait of a well-known politician, but it is fairly safe to assume that a ten second shot is long enough in most cases. Economics add an exception to the rule: a picture bought for a large sum because of its exclusivity or rarity value should be exploited fully, even though the picture itself (say, the last one of someone now missing) might be unremarkable. To dismiss this valuable property in a brief five or six seconds of air-time would be wasteful.

However long it is held on the screen, every still should be used to its maximum advantage by introducing it into the narrative at a point which helps to add emphasis to the story. It must not be allowed to 'drift' in, apparently at random, causing momentary but serious confusion for the viewer. The principle applies to the simplest script lines:

Referring to the ...

(Introduce still of Chancellor)

... latest rise in interest rates, the Chancellor said ...

Bringing in the still a few words later makes all the difference:

Referring to the latest trade figures ...

(Introduce still of the Chancellor)

... the Chancellor said ...

Choosing the right moment at which to return to the reader in vision is just as important. It is not acceptable to whisk the picture from under the viewer's nose without good reason. Much better to wait until the end of a sentence.

Where events call for a sequence of pictures, it is important to maintain the rhythm by keeping each on the screen for approximately the same duration. Six, five and seven seconds would probably be reasonable for three successive stills referring to the same subject, five, twelve and eight would not. The temptation to go back to the reader on camera for a few seconds between stills should be avoided, otherwise continuity is broken. In this context, a brief shot of the reader becomes another but unrelated picture, interrupting the flow. If returning to the reader during a sequence is unavoidable, it is far better to make the link a deliberately long one.

Writing to a series of stills can, and should, be extremely satisfying, particularly when action shots are involved. Unlike moving pictures, where the writer is sometimes a slave to editing grammar, stills may be arranged in any required order to suit the script, and even a fairly routine item is capable of being made to sound and look interesting:

Four men have been rescued from a small boat off the Kent coast. They'd been adrift for twenty-four hours after the boat's engine failed during a storm. Their distress call was answered by ...

(Still 1, helicopter landing)

... a Royal Air Force helicopter, which flew them forty miles to Dover for medical treatment.

(Still 2, man on stretcher)

The owner of the boat is now in hospital suffering from hypothermia: the three others have been allowed home.

(Still 3, boat in sea)

Attempts at salvage are thought unlikely before the weekend, when the forecast is for better weather.

Apart from the opening news point, the story is told and the pictures shown in chronological order for ease of explanation. Note, too, that the basic rule of television scriptwriting has been observed even in this short example: the commentary complements what the pictures show and does not merely repeat what viewers are able to see and hear for themselves.

Still 1 says where the rescued men were taken and for what purpose.

Still 2, while identifying the stretcher case and his condition, does not leave the viewer wondering about the fate of the others, of whom there are no pictures.

Still 3 completes the story by looking to the future.

From a production point of view it makes sense to go from one still to another by using the same technique – cutting or mixing. The use of an electronic wipe from one to another should be reserved for changes of subject. It is also easier on the eye to use either all colour or all monochrome pictures for the same sequence. If they have to be mixed for some reason, the two types should be shown in separate blocks.

Zooms or other movements should be used sparingly.

Composites and split screens

One step beyond the simple still comes the composite or split screen. In their most common form these consist of two heads, each occupying one half of the screen. For preference the faces should look slightly inwards towards each other, and not out of the screen, and they should be matched in size, style and picture quality, even if it means copying one or both originals, or manipulating them by computer. The commentary should always identify the characters from left to right.

A three-way split or more is also possible, but the screen tends to look cluttered. Reducing each picture to 'postage stamp' size on a plain background is not always effective, especially as identification is likely to take a disproportionate amount of time.

Other versions have part of the picture occupied by a still, the other by some form of artwork. Written quotations and maps can look particularly striking if illustrated in this way. Modern technology allows the words to be 'zipped' onto the screen separately but spoken commentary and text must match exactly.

The opportunities which now exist to employ such a wide range of styles are undeniable advantages to television news as part of its job to make programmes visually attractive as well as informative. Yet it is precisely because all manner of stills are available in such quantities that the danger of overkill is ever present.

Writers must avoid the temptation to use them automatically, almost as a reflex action, without thinking whether they are adding to the viewer's understanding. Seeing the screen littered with faces of the best-known personalities every time they are mentioned in a news programme quickly becomes irritating. The fact is that some stories are better told with no illustration other than the presenter's face.

It is more than thirty years since (Sir) David Frost made the cruel but accurate comment about the obsession of some television people to illustrate everything, no

Split screen. Uneven composition looks untidy and is bound to create a distraction for the viewer.

Split screen. The two pictures should be of similar size and quality, with the subjects looking toward each other. Identify the characters left to right.

Three way split. Names are printed to avoid confusion, and the commentary must follow in the same order.

Split screen quotes. For long quotations a series of captions can be used.

Split screen using a still and a map.

Figure 5.5

matter what the image might be. It became known as the Lord Privy Seal Syndrome. In Frost's television sketch the words '...Lord Privy Seal' (a Parliamentary title) were used in a news item with three still pictures over each word. It still stands as a warning to anyone lured into the use of pictures for their own sake.

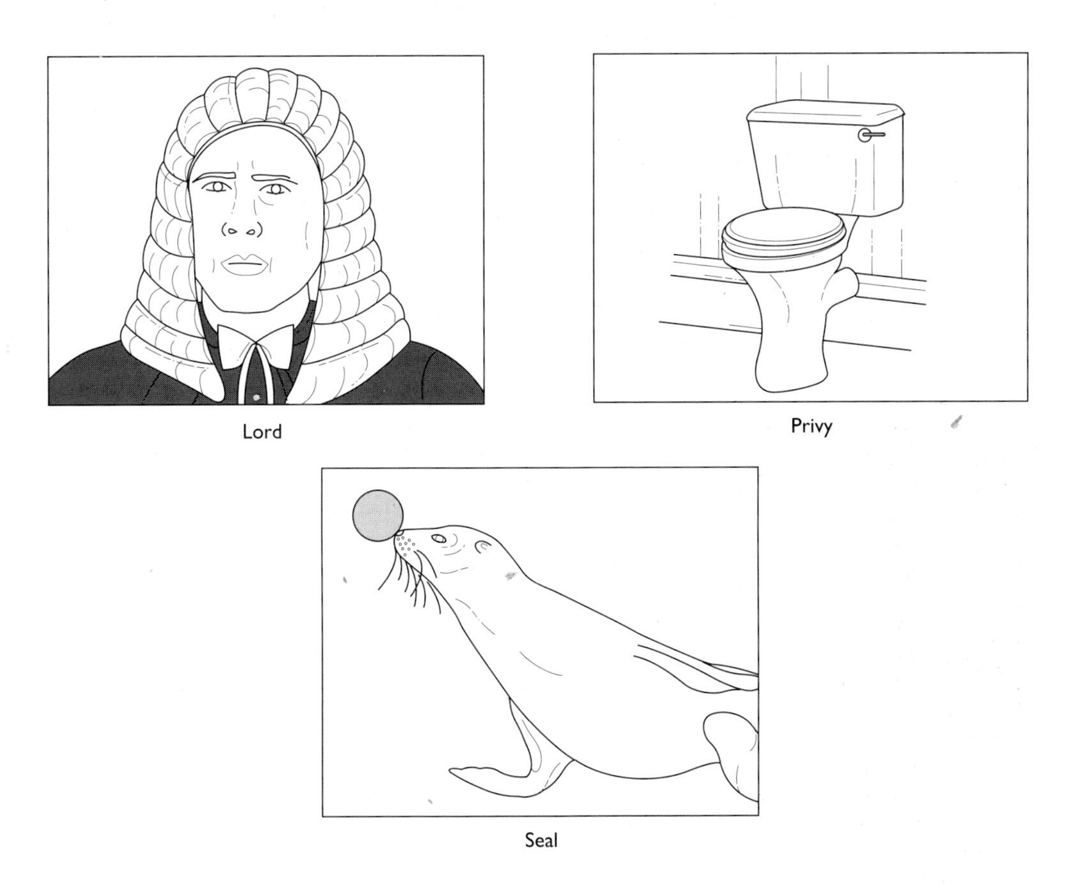

Figure 5.6

Electronic graphics

Complementing still photographs are graphics, which make use of some of the most powerful and fascinating software available in television and combine these with the journalists' need to convey complex information in an easily digestible way. Maps help the immediate identification of geographical locations; diagrams and charts allow sports results to be digested more quickly and expand on detail not easily understood when given by the reader alone; budget proposals, the main conclusions of official reports, detailed timetables or events, the ups and downs of interest rates and trade figures are made all the more palatable by the judicious addition of illustration. As important is the impression created in the mind of the viewer, who expects news to demonstrate the same high production values as the programmes around them, and graphics provide editors with excellent opportunities for changes of pace, variety and punctuation.

How artwork is prepared depends on available resources and the extent to which specialist staff and materials can be properly employed. Some very small or impoverished news services still have to get by with part-time artists, maybe staff members spared briefly from other duties, or design students happy to earn small sums for an hour or two's extra labour of love.

At the other end of the scale are the extensive graphics departments of the major players, whose belief in graphics as an essential weapon in their armoury has led to the investment of substantial sums in a panoply of big memory computers (usually with massive screens attached) and a staff of designers and artists to operate them. It is a clear case of artistic and computer skills coming together and that has now become a commonplace in art galleries as much as in television production. All manner of effects, including complicated animation and 3D, are routine in daily news programmes, but the creation of a computer graphic can sometimes be a slow and frustrating process. Complex fine art may call for special programming of computers or the transfer of images from video or other sources.

Probably the first and most influential of the basic electronic graphics systems was the Quantel Paintbox. This, as its name implies, allows the graphics designer to 'paint' coloured artwork directly onto the screen, although that description does not do justice to the range of facilities available. It is not necessary to know anything about either electronics or computers to use it to the full extent of its considerable capabilities, because it has been designed for use by a trained artist, not a graduate engineer, and the terms used are deliberately those with which an artist would be familiar.

The artist sits in front of a tablet representing a smooth, blank drawing board, and draws on it with an electronic pen which leaves no mark. Instead its movements are reproduced on a colour monitor screen, its position indicated by a small cross (cursor).

Figure 5.7 Paintbox graphics. (Photo courtesy of Quantel Limited.)

By moving the pen to the bottom edge of the tablet, the artist is able to display on the screen a palette of colours contained in small square 'paint pots', together with an area for mixing other shades and colours. When the cursor is positioned over one of these pots, a sharp tap on the pen lets the artist charge his or her 'brush' with the chosen colour which may then be used to 'paint' on the 'canvas.' Blends are achieved by applying pen pressure on the mixing area to deposit one colour, recharging the brush with a new one and depositing in the same way. If required, the artist can leave an inexhaustible supply of the newly mixed colour in one of the 'paint pots' and the whole palette can be stored electronically in the memory for later recall.

A panel to the right of the palette allows the choice of one of several brush thicknesses, again by tapping down on the appropriate square. A menu of other options allows for the selection of an airbrush, for example, a faithful reproduction of its texture being shown on the screen.

Another menu option selects 'graphics', which provide for the drawing of separate or connected straight lines by 'rubber-banding'. One tap of the pen positions the start and a second tap the finish, the line being stretched between the two points. Using the same method, shapes can be drawn to any size, outlined or filled, anywhere on the screen, in any colour, thickness or texture.

The paintbox, now an accepted generic term, also provides electronic versions of stencilling, cutting-out and pasting used in conventional graphics. In stencil mode, a mask is made to protect one part of the picture while work is carried out on the rest using all of the machine's facilities. Further stencils can be generated within the area being worked on, and any part of the picture covered by the stencil can be revealed and the remainder protected. Every stage can be stored in the memory, so if the next process does not give the desired effect nothing is lost, as a perfect copy of the previous artwork is instantly available.

The artist can start from a blank canvas primed to any chosen colour or a background shaded from one colour to any other. A still picture or a frozen video frame can be worked on. Blemishes on faces can be painted out using an exactly matching colour picked up on the brush from elsewhere on the 'skin'.

Any part of the picture can be cut out electronically without destroying the original and then enlarged, reduced, turned over top to bottom, reversed left to right, pasted anywhere on the screen, etc., processes which can be repeated as often as required. The same effect can be achieved with different styles of type, which can be cut, manipulated and then pasted into any position in any colour, outline or solid, plain, embossed or with drop-shadow (of any depth in any direction).

Library storage and retrieval, with comprehensive searching and browsing facilities, represent only some of the capabilities of the paintbox. New features are being added all the time. It is a way of producing graphic work for television of a high technical quality and at the same time giving the artist freedom of expression. Probably the only drawback is that the cost of the equipment is likely to mean that few television news graphics departments can afford to buy a paintbox for each member of the team.

Since the arrival of the paintbox, advances in technology have brought with them the development of other equipment, including the digital video effects (DVE) generator of which Charisma, also manufactured by Quantel, is an example. Charisma is able to create movement and manipulate images. Harriet, a combination paintbox, videotape and digital effects machine, is particularly effective at incorporating videotape into a graphics sequence. Among other leading manufacturers are Abekas, whose range includes a digital disk recorder which takes in information and images from different sources to build up multi-layer graphics without loss of quality, and Vertigo, with a 3D system capable of modelling and animation.

Another important advance has been the creation of off-line graphics areas, where the recording of particularly complex graphics or sequences can be undertaken for use at a later stage.

The artist and the writer

In the past newswriters did not necessarily expect to be overly concerned with the fundamentals of graphics design of any sort. But now that the digital age has added this new dimension to television art, interest in and knowledge of what is possible has become integral to the writer's responsibility. Helping to create exactly the illustration necessary to accompany a news item has become part of the job.

Not surprisingly, graphics designers often develop love-affairs with their exceptional toys and what they are able to create, and it is up to the journalist to ensure that their enthusiasm does not carry them away. Graphics for the sake of graphics are no help to the viewer. Having said that, the opportunity for experimentation is almost endless. The imaginative use of graphics can bring life to almost any routine news item, while big news stories often lend themselves to dramatic treatment. Among international events the Gulf War and the conflict in Chechnya stand out as offering opportunities for the creation of special effects to explain what happened in a news event for which pictures are not available.

Most artwork for television news has a screen life of no more than a few seconds, and for that reason if nothing else, it must be clear and easy to follow, with bold images on unambiguous backgrounds. An awareness of the dimensional limitations of the screen also helps. Among the most important of these is the amount of information that can be squeezed in, bearing in mind that unnecessary clutter reduces visual impact and 'cut-off' (the edge of the television picture automatically lost in the process of transmission) reduces the area the artist has to work on. The effect is exaggerated on old or misaligned receivers, so some of the information intended for the audience is lost. The hazards are trying to squash too much into a limited width so exaggerated that cut-off mutilates each end of the lettering, or making the information occupy so much room that part of the background illustration itself is obscured.

(a)

(b)

Figure 5.8 (a) Too much lettering obscures part of the face. (b) Abbreviate the title. Smaller lettering would be hard to read.

Above all, using the designer's skill to illustrate a story must anticipate how the accompanying commentary will be structured. There is little point in commissioning good-looking charts full of important details to which no reference is made, or in cramming beautifully scaled maps with place names which the script ignores, to the confusion of the viewer, who is then left wondering why they were put there in the first place.

The temptation to add anything except the strictly relevant must be resisted. In the report of our Pitkin's Bank raid, a map illustrating the location of Luton needs only the additional reference points of London and the most important trunk road, the M1, to

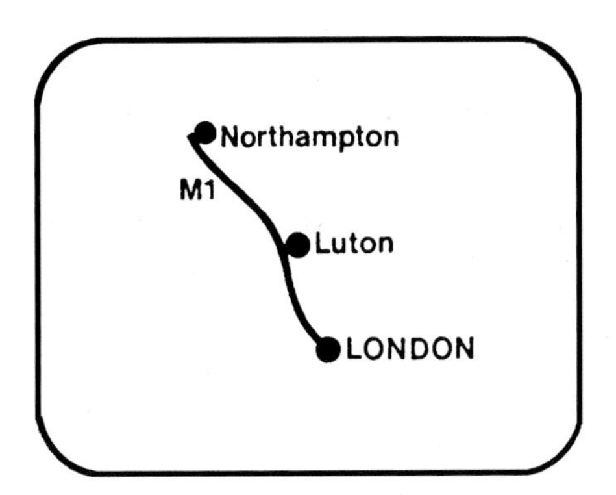

Figure 5.9 (a) Enough detail is necessary to give the viewer a point of reference during the few seconds the map is on the screen.

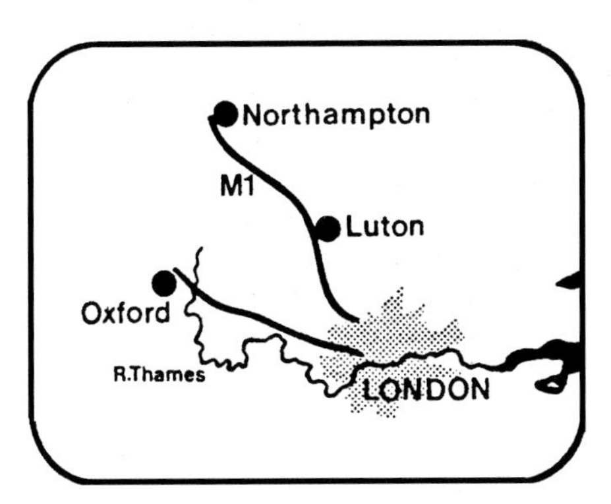

Figure 5.9(b) A prettier map, but the extra place names will only confuse the viewer unless the commentary makes some reference to them.

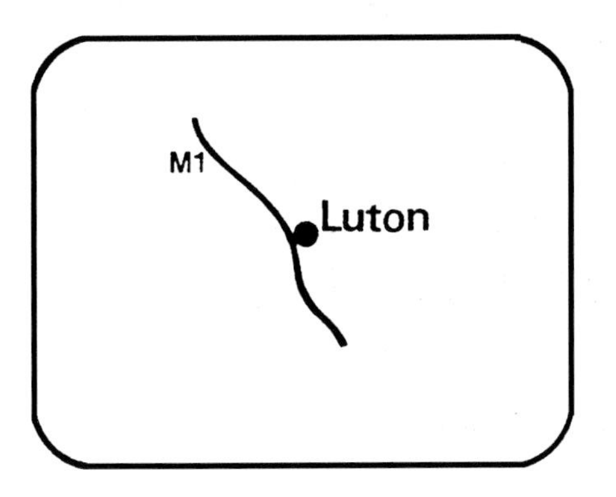

Figure 5.9(c) The other extreme. No reference point means the audience will have no idea how the place central to the story relates to anywhere else.

give a clear idea of the town's location in southern England. The arbitrary addition of several other big cities and county names may make the map look more attractive to some eyes, but would do nothing for the viewer during its few seconds on the screen. At the other extreme, a large-scale map isolating Luton from everywhere else in the country would be equally valueless.

The key to success in writing to graphics lies in the assembly of words in sequences which lead the viewer's eyes progressively across the screen from left to right, top to bottom. Anything which gets in the way is bound to be distracting, so any movement within a graphic should be kept to a minimum, especially if it has to compete with text.

DEFENCE SPENDING

	Then	Now
Tanks	£300m	£375m
Ships	£670m	£805m
Planes	£795m	£990m

Figure 5.10 Simple charts with or without suitable backgrounds help to get figures across. The accompanying commentary must lead the viewer's eye naturally from top to bottom, left to right.

Younger viewers, with senses attuned to fast-moving computer games, are probably able to cope with it all, but for the sake of the general audience there is much to be said for simplicity. Not that it is enough for the reader to repeat what the viewer is seeing. Some complementary information must be fed in as well, but if, through carelessness, too wide a difference is allowed to develop between what the audience is being told and being shown, the effect that writer and artist have been at pains to create will be destroyed.

That is likely with even the simplest illustrations:

AVERAGE COUNCIL TAX

North UP £80	From April, an average of ninety-eight pounds goes on bills in the south, bringing the year's figure to four hundred and fifty-three. An average increase of eighty pounds makes the total in the north four hundred and ten.
South UP £98	

The writer's mistake here is virtually to ignore what is on the chart: there is simply no correlation between what the viewer is seeing and hearing. Transposing the two sentences will help a little, but not much.

AVERAGE COUNCIL TAX

North UP £80	From April, an average of eighty pounds goes on bills in the north, bringing the year's figure to four hundred and ten. An average increase of ninety-eight makes the total in the south four hundred and fifty-three.
South UP £98	

It would be far more effective to use the same phraseology as on the chart, and in the same order. Following the 'left-to-right, top-to-bottom' guideline would bring about a clearer message:

AVERAGE COUNCIL TAX

North UP £80	From April, tax bills in the north will go up by an average of eighty pounds, bringing the year's total to four hundred and ten. In the south, an average increase of ninety-eight pounds will make the figure four hundred and fifty-three.
South UP £98	

Probably the most admired exponents of the art are the makers and writers of television commercials, who have very similar periods of screen time in which to get their message across, and need to establish immediately unambiguous links between what the viewer is seeing and hearing.

The technique is all the more important where information is added one stage at a time. This, always a popular way of emphasizing facts or figures, has become even more common since the introduction of electronic equipment. The effect is achieved either as part of graphic design, or by a superimposition (mixing two sources to produce a single picture) at the appropriate moment during transmission.

In the case of our Council Tax graphic, the first spoken words would accompany one fact already on the chart:

AVERAGE COUNCIL TAX

North UP £80	From April, tax bills in the north will go up by an average of eighty pounds, bringing the year's total to four hundred and ten. In the ...

The next part of the commentary would be spoken simultaneously with the introduction of the stage completing the picture:

AVERAGE COUNCIL TAX

North UP £80	... south, an average increase of ninety-eight pounds will make the figure four hundred and fifty-three.
South UP £98	

Elegantly designed charts and diagrams illustrating the ups and downs of stock markets, interest and currency exchange rates and other statistics have become among the most familiar sights to viewers of modern television newscasts, but in commissioning them writers must take note of the need for accuracy. Comparison of figures must take account of the scale of the graphic, so that, for example, the impression of a huge variation from one month to another is not given when in fact the change is only minor.

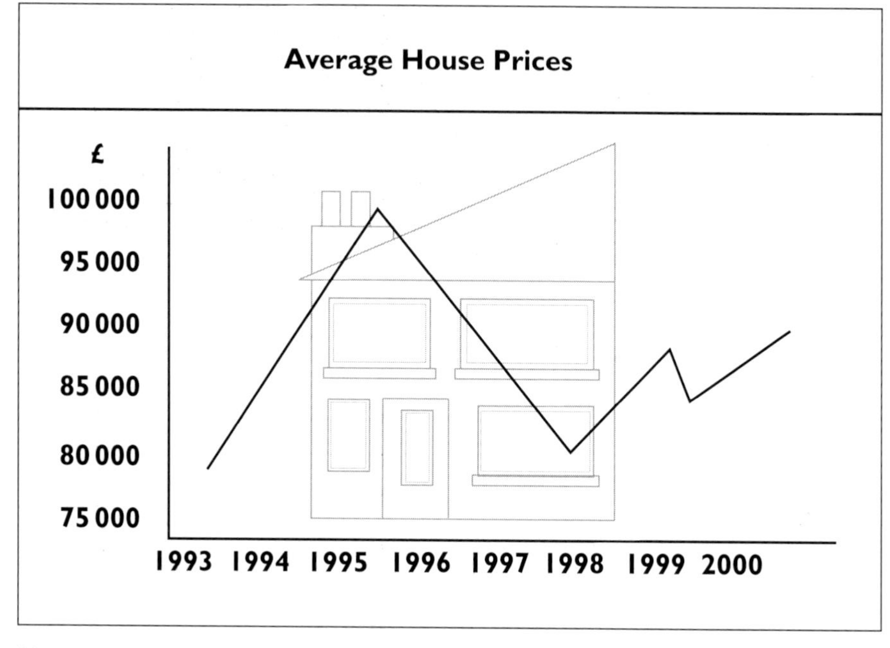

(a)

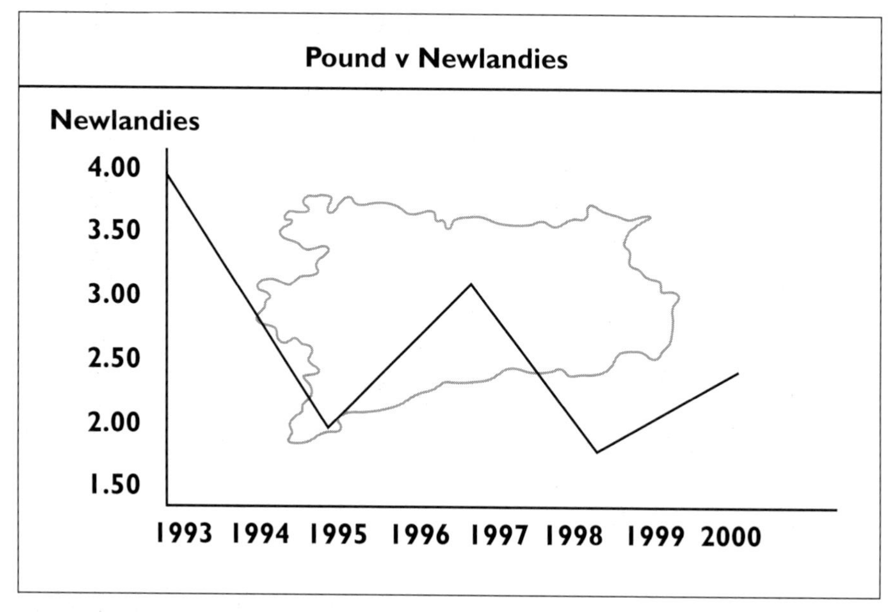

(b)

Figure 5.11 Charts and graphs must show details in proportion. The imagined fluctuations in 'Average house prices' in (a) are represented quite effectively, while the base line in (b) is too low, making the pound's value against the mythical 'Newlandies' appear to swing more wildly than necessary.

(a)

(b)

(c)

Figure 5.12 Electronically keyed insets are sometimes used to accompany presenters for stylistic purposes, or to enliven stories read in the studio. Items can be enhanced by well-proportioned, appropriate illustrations (a) but viewers may be distracted by those which are poorly chosen or positioned in a way that dominates the reader (b). The addition of key words to insets is also a useful device (c) as long as the wording is not so obvious as to patronize the viewer.

Another very popular form of graphic, known as the DOG (digitally originated graphic) or inset, appears behind or beside a studio presenter as part of programme window dressing. Size, shape and position on the screen are usually considered very carefully, but the impression is sometimes given that not nearly sufficient thought or care goes into their design or content. Some seem to have no bearing on the story they are meant to illustrate; inappropriate inset stills dominate the picture or appear to be threatening the reader over his or her shoulder; composite graphics, otherwise well-produced, include text which may be too difficult to read or so simplistically descriptive as to patronize the viewer. Worse still, spelling mistakes are not uncommon. Hard though it may be to accept, the responsibility for proof-reading lies with the writer. Completed artwork must be checked carefully against any original plans, however sketchy. Where sequences of graphics are involved, no doubts must be left about the exact order in which they are to be transmitted.

To have one illustration on the screen while the commentary is clearly referring to another is hardly likely to inspire the audience's confidence. Misspellings, which somehow seem to occur in only the most simple, everyday words, have a tendency to harvest bulging postbags of complaint. It is true that genuine variations do exist in the spelling of certain place names, particularly where they are transliterations, and news agency reports frequently contain spellings which differ widely from those in atlases or gazetteers. To the domestic audience it may scarcely matter as long as there is consistency, and any confusion created by changes in common usage, for example the switch from Peking to Beijing, dispelled as quickly as possible.

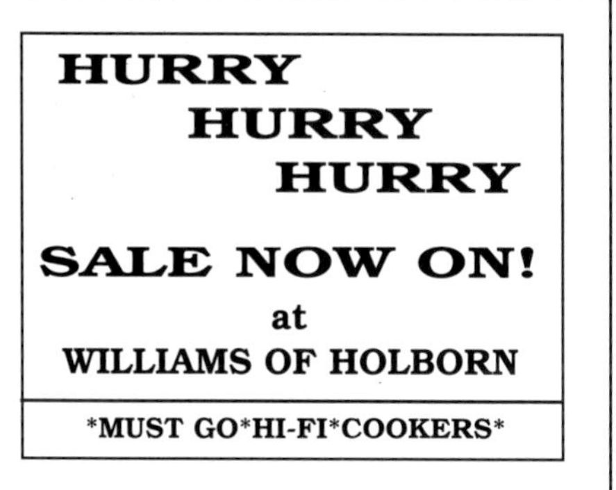

Premier League			
Arsenal	4	Sunderland	1
Chelsea	1	Leicester	1
Coventry	2	Wimbledon	0
Everton	2	Tottenham	2
Middlesborough	1	Derby	4
Sheffield Wed	2	Bradford	0
West Ham	1	Aston Villa	1

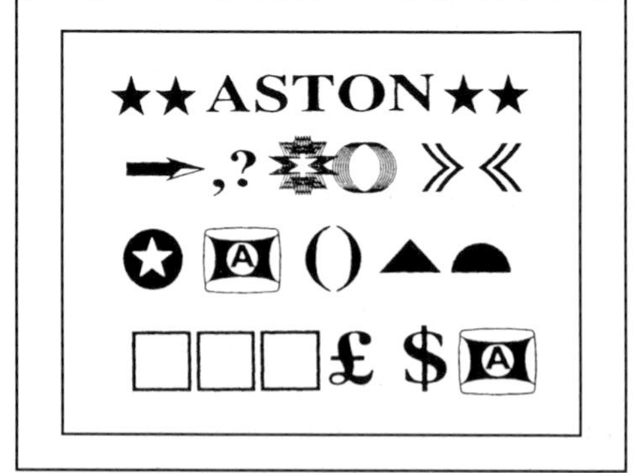

Figure 5.13 A wide range of typefaces can be created electronically in several sizes as text on charts and for subtitling. Foreign-language versions are also available.

The solution to all these problems is to accept certain standard reference books (*The Times Atlas*, the *Oxford Dictionary*, etc.) as the arbiters and ignore everything else.

Pictures with tape sound

So far the emphasis has been on the addition of visual material to accompany words spoken by the presenter in the studio. Now comes the moment to create a temporary but complete replacement for the reader by the use of sound-only reports. These are the blood brothers of the dispatches which have become the stock-in-trade of radio news programmes all over the world. For use on television, telephoned sound reports, whether 'live' or recorded, come a poor second to moving pictures or other forms of illustration. But the nature of world news demands that the voice of the reporter on the spot is, on many occasions, infinitely preferable to nothing at all. What it usually looks like on the screen is a simple composite graphic of either the reporter's face, or the place he or she is reporting from, and a sound-only report. Only the news value of the story dictates whether to run this on television. This kind of quick sound-only snatch of news became unfashionable in the late 1990s and in some cases many news programmes faced with a 'while-we've-been-on-the-air' breaking story will simply prefer to do a two-way (interview)

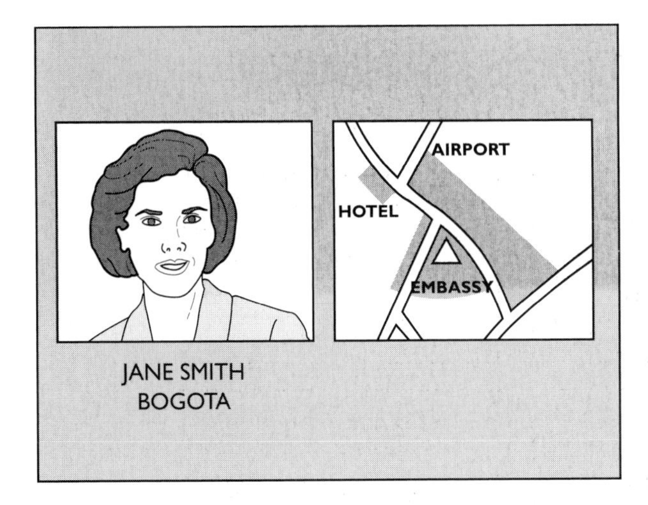

Figure 5.14 A composite graphic to cover a voice-only report usually identifies the correspondent and location. The main difficulty arises when a dispatch is made from one place about events taking place in another.

Widescreen transmission

Britain, and the BBC in particular, is ahead of most countries in transmission of material designed for a widescreen TV set. A traditional TV picture shape is four units long and three units high, called 4 × 3. Widescreen television is more like the shape we see at cinemas. This is sixteen units long and nine units high, or 16 × 9. This means it is wider than traditional television. Several news services are *transmitting* on widescreen, so the problem is that the usual pictures brought into the news operation simply do not fit when they go out on transmission. They can be broadcast, but for the viewer they look odd, with a stream of distorted, squashed or thick human bodies moving about the screen and the viewers left wondering whether to adjust their sets. The solution is an aspect ratio converter (ARC). This piece of equipment will re-size a picture, like placing a round pin into a square hole. It converts standard 4 × 3 ratio, such as shot by a standard news camera, into widescreen 16 × 9. New material as well as archive can be put through the ARCing process. Widescreen offers the viewer more of a picture, quite literally. It also has greater scope for combining several windows of information on screen at the same time – the presenter, graphics, lines of information moving in another corner and, on television sets capable of Internet access, a view of associated websites.

Figure 5.15 The widescreen television picture, based on an aspect ratio of 16 × 9, enables the screen to contain much more information. This is the BBC's continuous news service, BBC News 24. At one moment the presenter can be filling the screen and then be moved into a 'box' at one side as graphic information and numbers are added to the screen. It is particularly useful in this case for business news. Widescreen TV sets are expected to dominate the market by the end of the decade.

Figure 5.16 The viewer has become more accustomed to the idea of shapes which move and shrink or expand. For example, this composite interview shot in which participants do not fill the screen.

6

Words and images

The power to transmit moving pictures no longer exclusively sets television apart from its rivals in the business of disseminating news about events of importance and interest. Images available across the web improved in definition and clarity throughout the late 1990s. All media have become fused in a kind of symbiotic wedding. Television, radio, newspapers, magazines and the web don't always get along, but they do need each other to fuse new commercial technology in the decade beyond 2000.

For our purposes we know that Content is All when it comes to transmitting television news. The skills to write a script that works with moving images in any medium are the same. For the newcomer embarking on the acquisition of basic skills to cope with this extra dimension, two major hurdles bar the way. First and most important is the need to develop an instinct for the construction of written commentaries in a way which allows the viewer to draw full value from both words and pictures. Second comes the requirement for at least a rudimentary working knowledge of the technical tools which provide the means to this end.

The lasting influence of an older word – film – is not to be overlooked. By the word film we do not mean a 'movie'. Film once breathed life into the new form of journalism called television news and set the pattern for what we have now. Producers who may never have used it still talk about 'films' and 'film-making' because – in Britain at least – the terms 'taping', 'videoing' and 'DV [digital video] capturing' have not widely caught on in the industry.

Film, electronic video, digital tape and disk have their own champions lauding texture and quality, speed and flexibility, and enough has been written about them to keep the keen student occupied for years. But it is important not to lose sight of the fact that while these media use different methods of storing the image, the art of picture-making is common to all.

It is not necessary for every writer to make a full-time study of the subject, but it is clearly important to understand the moving picture as a device which can be employed with great delicacy, even within the large blunt instrument which is the routine television news programme. Many television journalists whose ingenuity and imagination in the use of pictures have outgrown the confines of news have gone on to find success in other areas of television which combine subtlety with qualities of journalistic hard-headedness.

Perspective: the history of the lightweight revolution

The first breakthrough towards mobility in collecting sound and vision came in the 1950s. Until that time the experts were dismissive of anything smaller than the 35 mm film format used for cinema features and what were called 'newsreels'. Newsreel camera-operators did a brilliantly courageous job during wartime, and the quality of the images they produced was undoubtedly superb, but the equipment for sound recording and filming was cumbersome and virtually impossible to use without a sturdy tripod. This partly explains why in so many of those historic film interviews which are shown on television from time to time the subject has clearly been brought to the camera and microphone, emphasizing the air of stiff formality and self-consciousness that was always so apparent.

The introduction of 16 mm film – by, it is said, American news executives who wanted to find a better way of covering the Korean war – at last made it easier to take the camera closer to the subject. Sixteen millimetre was a format familiar mostly to amateur film-makers. The camera was still heavy, but at least it could be balanced with reasonable comfort on a shoulder or screwed to a monopod and carried into action. The sound was recorded onto a thin strip of magnetic tape attached to one edge of the film during manufacture.

One award-winning veteran camera-operator used to like recalling how the makers of the first 16 mm camera he was issued with were so sure it was for amateur use that the instruction booklet included something along the following lines: 'Having placed the camera on the tripod, seat the subject at the piano, making sure you are at least five feet away and that sufficient light is available from the window.' So equipped, he and the rest of the world's television news camera teams went to war to bring back vivid pictures which made the viewer feel personally involved as never before. He survived many of the conflicts of the late twentieth-century: Vietnam, Biafra, Aden, Cyprus, the Middle East, Northern Ireland and many other trouble spots, only to be killed one night as he crossed the road near his home.

For about twenty years there was nothing to touch 16 mm film. The cameras were reliable and impervious to most of the knocks and other ill-treatment they inevitably suffered in the field. The film they consumed – first monochrome, then colour – was fast and versatile.

Some news organizations did experiment with a format called Super 8, commonly used for home movies, as a way of avoiding problems at airports where the arrival of news camera teams and their unmistakable silver boxes often led to bureaucratic delay or the impounding of equipment by unsympathetic customs officials. So there were some occasions when a camera-operator who would have been denied entry with an Arriflex or Cine-Voice was allowed in on a tourist visa and contrived to shoot a story on what was regarded as a 'toy' camera.

But even if no serious attempt was ever made to replace 16 mm with 8 mm, a complete alternative to the use of any film for straight news work was becoming both an attractive and practical proposition.

Film had its limitations. Until processing was completed, not even the world's greatest camera-operators could guarantee that they had measured the light correctly, there was not some other fault, or that the images they believed they had captured were indeed on the film in the way they had intended. And since news by its nature happens only once, there was never the possibility of going back to re-shoot it, unlike feature film work, where expensively assembled casts and technicians are kept together until the camera-operator's efforts have been seen and approved. To add to these hazards, despite the best endeavours of the news laboratory workers and their increasingly sophisticated developing equipment, film sometimes did 'go down in soup' (chemical processing) to be lost for ever.

The biggest drawback of all was that even though processing time was being reduced to the point where it took no more than a few minutes to develop each 30 metre length of film, it was still comparatively slow. In news, time is a luxury which can rarely be afforded.

It was for that single reason that television news and some other topical television programmes had already cut out many of the intermediate stages in conventional film-making, so that progress from camera to screen would be as short as possible. Instead of first making prints from an original negative, they took the developed negative material and transmitted that. In adopting this method, television news people had to accept the very real risk that a careless or unlucky film editor, working under deadline pressure with well-worn equipment, might do irreparable harm to precious material, and that in some circumstances an edit once made might be impossible to restore. That in turn put greater responsibility on the editorial staff to ensure the right decisions about editing the film were made first time.

To compound the difficulties, film editors and writers working for stations transmitting only black and white pictures had to make their decisions based on the identification of people and events from film viewed in negative, as delivered by the processing department. After a while, experience taught that it was possible to identify the better-known public figures. But quite frequently an amount of guesswork had to be employed, not always with complete success. Fortunately, the viewer was spared such uncertainties. During the film's transmission stage all was put right by the use of 'phase reversal', an electronic means of changing the blacks to white and the white to black, thus producing a normal positive image.

Introducing colour

Much to the relief of every writer, the problems associated with identification disappeared during the switch from mono to colour filming, which began in the 1960s. After much discussion and experiment many news services opted for a particular 'reversal' film stock, which meant that once it had been through the processing bath it appeared in the same form as the amateur photographer's colour transparency, as a positive which was taken directly to the cutting room for editing for transmission.

During the same period developments in recording pictures and sound on to magnetic tape were taking place almost in parallel. The system may not have been originally conceived as one which would necessarily benefit television news, but it seems significant now that the very first broadcast using videotape, on 30 November 1956, was that of Douglas Edwards and the news which was transmitted by CBS from New York, recorded in Los Angeles, and replayed three hours later for viewers on the West Coast of America.

It had long been recognized that some form of recording system for television was extremely desirable, if only to give studios the flexibility to create their productions at times which suited them and their participants rather than forcing them to continue taking all the risks associated with transmitting programmes live. One step towards that end was tele-recording, a way of filming productions or production segments off high-quality monitors. Despite its success, this system had to rely on the conventional photographic technique of chemical processing, with the inevitable delay before recordings could be examined.

The main snag in developing instant playback along the lines of quarter-inch tape sound recordings lay in the very high speed and consequent amount of tape expended to reproduce pictures of sufficiently high quality for broadcasting. The problem was eventually solved by Ampex, an American company which had begun its research into magnetic tape recording for television as early as 1951. By early 1956 the company's engineers were able to demonstrate a machine on which the speed of a 2 in (51 mm) wide tape was kept down to 15 in (38 mm) a second as it moved past four recording heads rotating about a hundred times faster.

The end result of this quadruplex technique was picture reproduction which most lay people found indistinguishable from the original, and for the name of Ampex a permanent place in the language of television.

Television news, for which perfect picture quality had never been the priority where important news stories are concerned, believed the biggest advantage of video was that it could be replayed in no greater time than it took to rewind the tape, opening up many previously undreamed of possibilities for flexibility within programmes. Many news items which began their lives in the 16 mm sound camera now ended them on the screen as videotape recordings, for film which could be processed and edited at regional or other television stations was then available to be pumped along public telecommunications cables to headquarters. In addition, anything photographed by the electronic cameras in the studio or outside could easily be linked to a videotape recorder. Coupled with the strides then being made in intercontinental communications systems, the development of videotape suddenly opened a new era for the gathering of foreign news in particular, and the whole effect was to extend deadlines until the ends of programmes.

The equipment remained expensive to buy and install, particularly when colour came in. The cost was only partly offset by the fact that recordings could be erased and the same 90 minute tape used over and over again until 'pile-up' (thick horizontal lines of interference) showed its useful life was over. Another disadvantage was the size of the machinery. Each quadruplex resembled an overgrown reel-to-reel audio recorder, and together with its ancillary equipment took up the space of a small room. Studio and outside broadcast cameras were even less manoeuvrable than the old 35 mm film equipment.

It was not until the 1970s that real progress was made towards producing a tape which was smaller yet still of a standard high enough to satisfy the exacting requirements of the broadcast engineers. When it came it was of a single size, 1 in (25 mm), and in two technical specifications, because two solutions had been found to the problem of cramming the same amount of audio and video information onto a tape surface half that of the original Ampex.

The appearance of 1 in for studio-based work was followed swiftly by recorders and cameras which allowed more than a degree of portability, but the start of the true lightweight revolution probably dates from the time American broadcasters began experimenting with Sony U-matic, a Japanese system which had originally been designed for industrial use.

The biggest advantage of this format was that it did not use the tape spools needed for one inch: instead, sound and picture were recorded on to a three-quarter inch (19 mm) tape safely enclosed inside a strong plastic cassette. No handling of the tape was necessary. In the field the cassette was automatically threaded when inserted into a recorder which was carried in a leather case and slung over a shoulder.

The rest, as they say, is history. Once Sony had produced a new specification to satisfy the 625-line picture standard used by most of Europe, the way was clear – technically, anyway – for the biggest advance in news-gathering television had seen for 20 years.

The rise of ENG – electronic news-gathering

The first television station to take the plunge and replace its entire film equipment with ENG is believed to be KMOX-TV in St Louis, Missouri. Cameras, processing, viewing and editing facilities were all disposed of in one grand gesture in September 1974. Interest in the system was already spreading among other local American stations which were beginning to appreciate how ideally electronic cameras were suited to their needs for early-evening news programmes.

By 1976 ENG had dominated television reporting of the US Democratic and Republican party conventions and the US presidential election campaign. It was also being used in Europe: Japanese coverage of the London economic summit of May 1977 included ENG pictures which were beamed back to Tokyo by satellite and, as a bonus, transmitted by the BBC as part of their domestic output. But the European services themselves were moving more cautiously, opting for trial periods in which to evaluate the technical and editorial problems, while managements sorted out the changes necessary in staffing and retraining if ENG were to become permanent. NOS of Holland started their experiment using film as a back-up.

The BBC opened their 12-month trial on 10 October 1977, when an interview with Margaret Thatcher, then leader of the Conservative opposition, was recorded at the House of Commons and shown on the lunchtime news. ITN, at that time their only British television news rival, began their experiment during the general election campaign of April–May 1979, a year during which it was reckoned that more than half the television stations in the United States already had some ENG capability and more than 300 had followed KMOX-TV and gone all-electronic.

The final adoption of ENG elsewhere was still by no means automatic. Negotiations with the broadcasting unions were proving particularly difficult and long-drawn-out.

The networks had no doubts about ENG and committed themselves heavily. Within the United States it meant that facilities for processing 16 mm film were rapidly becoming so scarce that foreign-based news services faced the unattractive alternative of either airfreighting their undeveloped material home (with the obvious risk of it being well out of date before it reached the other side of the Atlantic) or buying American electronic coverage off the shelf.

Abroad, ENG was being used to the extent that the editor of one British television news organization complained that he would be able to use material sent by satellite from London and back by an American crew many minutes before film of the same story shot by his own crew on their own territory was out of the processing bath. To be beaten in his own backyard, he believed, was unacceptable. His argument may not have been the clincher, but by 1980 any remaining opposition was disintegrating, and the switch from film to ENG was going ahead at full speed.

How ENG works

The basis of electronic news-gathering is a lightweight camera and an integrated, detachable or separate video cassette recorder by which picture and sound are captured. Most systems are capable of being operated by one or two people, depending on staffing policy and story circumstances, but the trend is strongly towards the single-piece camcorder (camera-recorder) worked by one person.

Developments by 2000 had been influenced by a progressive reduction in equipment size and weight, together with improved quality, reliability and ease of use. Alongside

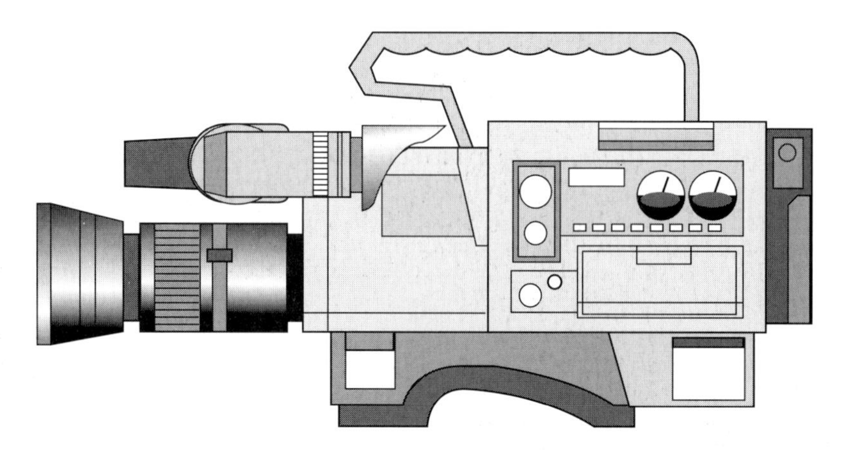

Figure 6.1

Beta SP (superior performance) and the camera/half-inch (12.5 mm) tape system used widely since the late 1980s, came DVC, the digital video system now more widely accepted. For the broadcasters it has offered economy and acceptable quality. There is a problem during the shooting stage when human ergonomics and physics mean the very lightness of the camera makes static shots more difficult to obtain. It is hard to keep the shot steady when the camera has such little weight. The biggest change of all now anticipates the replacement of tape altogether, with news-gathering on disk and in some cases on a disk removed from a disposable camera (the camera being no more than a cheap enclosure which holds the disk).

The advantages over film seem obvious now. No more delay while film is processed means the pushing back of deadlines: ENG teams can spend longer on the same story, or simply cover more. Coverage can (usually) be checked for quality and content on the spot. Electronic and digital tapes are cheap and reusable.

Most important of all is the flexibility the system offers. At the most basic level, an ENG or digital camera-operator or reporter on location records sound and pictures onto industry-standard video cassettes and returns them to headquarters by hand for editing and broadcasting. Where tape really comes into its own is when signals from the unit are fed back to base by microwave or even secure Internet link, live or already recorded. Some ENG teams travel in customized vehicles carrying their own equipment from which to transmit the pictures home. Others drive unmarked camera cars and have their transmitters installed in separate vehicles (sometimes known as facilities units, fast response vehicles or live links units) with which they can rendezvous at some convenient time and place. The tape is replayed to base by the unit's own cassette recorder.

As a much-used alternative, the camera signal is relayed from location to the mobile unit over a short-range radio link. Here again the pictures are recorded on the spot or simply bounced onwards using the aerial on the vehicle roof. (If the location is close by, the short-range link is in itself powerful enough to transmit the signal to base.) The aerial is aligned with microwave link dishes sited on the tops of tall buildings or masts which in turn pass the signal to receivers at news headquarters. Another option is to play the tape direct from the telecommunications points dotted around the country.

Many radio stations also have these 'plug-in' or 'inject' points from which ENG pictures can be played, and other permanent circuits are installed in urban centres.

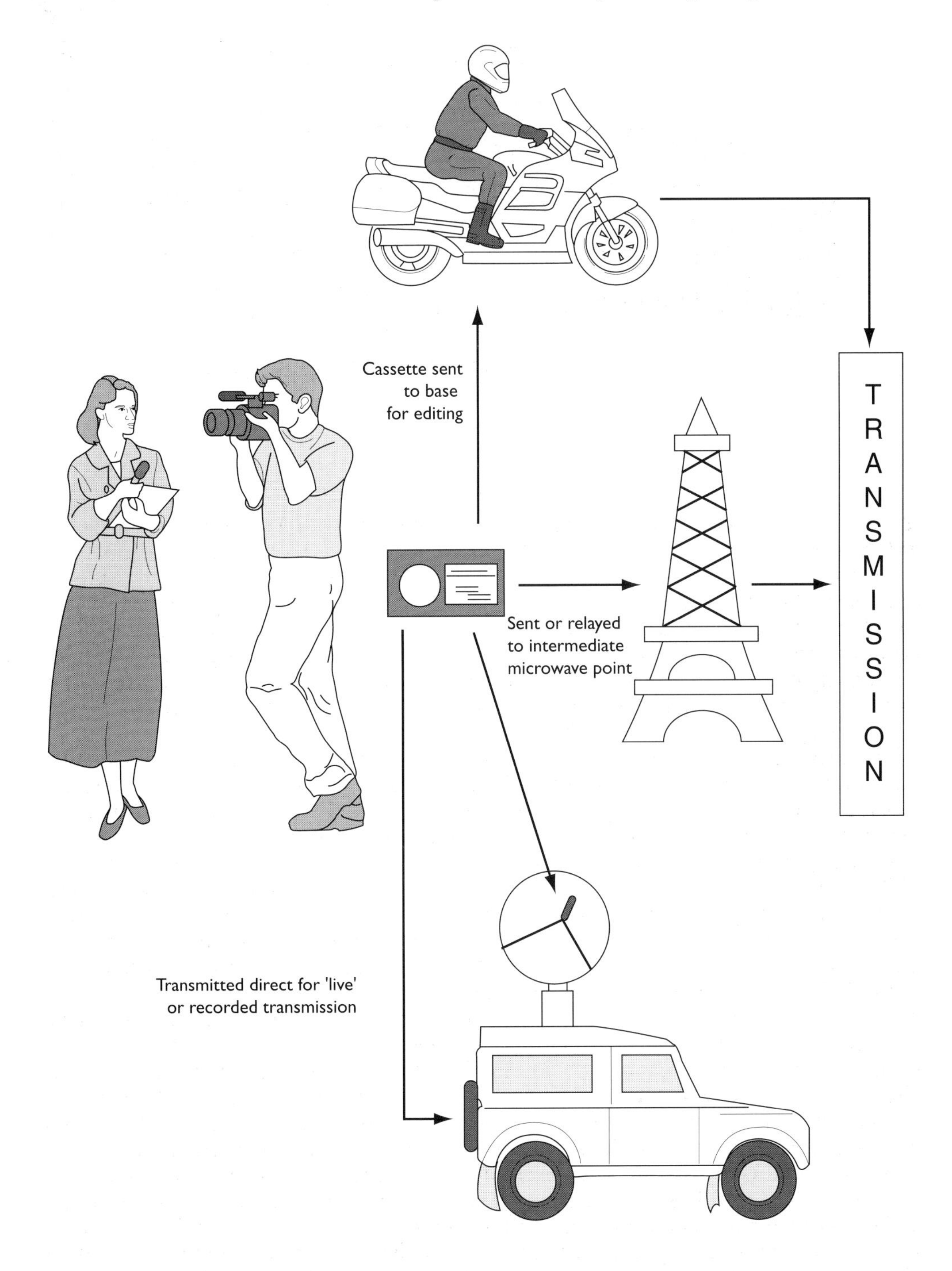

Figure 6.2 Material can be recorded on the spot then sent or delivered by hand or transmitted back to base directly or via an intermediate microwave link point. It can also be recorded and transmitted live simultaneously.

London, for example, has a sound and vision ring-main system which was originally established to feed outside broadcast pictures of the coronation of Queen Elizabeth II in 1953. These metal boxes are still sited near buildings at various places throughout the capital. When they are needed, the engineers open them, plug in sound and video leads, and route them by coaxial cable to the studio.

In very remote areas, where none of these systems exists or is near enough to be useful, an ENG team carrying its own mobile antenna and tracking equipment is able to transmit its material by bouncing the signal off a communications satellite. Satellite news-gathering (SNG) is now an integral element of news coverage in domestic as well as foreign locations, and without it the coverage of the Gulf War and the fighting in Bosnia and Kosovo during the 1990s would have been limited (see p. 152).

Picture editing

In an ideal world, every journalist would be able to sit at ease and watch the raw rushes from the camera on a screen big enough to appreciate all the reporting team has accomplished. Unfortunately, on most occasions, writer and editor go straight into an editing booth with their cassettes and find themselves working against the clock, able only to make instinctive decisions about material they may have time to view only once – and that at faster than normal speed.

What they select depends on the nature of the subject, its interest, importance and nominated position within the programme they are working towards. Equally important on occasion is the extent to which the two have briefed themselves, so by the time the tapes reach them they are already aware of the contents and have planned a rough scheme for editing. (Where the writers are also the reporters/producers this knowledge must be assumed.)

At this stage there is no room for clashes of interest or temperament – only teamwork to ensure that what appears on the screen, perhaps only a few minutes later, reflects the successful fusion of separate professional skills. In recent years this has become more possible than ever before, as relationships between picture editors and journalists are less strained than they once were. Most of the old-style cutters who originally left the film industry to join the pioneers of television news have now departed. With them has gone much of the friction which arose with writers who felt that getting the story across was always more important than sticking to some of the more rigid rules of film-editing grammar.

By the time the lightweight revolution had been completed, the old school had been replaced by a new breed of young editors who quickly came to acknowledge that they and the writers shared one common aim: to tell the story in pictures and words, as coherently as possible. The result is modern, streamlined and effective. Not that the younger picture editors are any more keen than their predecessors to break the rules. It is just that for the sake of simplicity they are prepared to dispense with the irrelevant.

It can also be considered fortunate that a few of today's more experienced editors were still in the early stages of their careers as videotape was taking hold, and so retained a precious knowledge of film techniques. Others who once spent their time working with the early videotape-machine monsters brought with them into the ENG editing suites hard-won technical knowledge which made it easier for them to make the transition to smaller format working.

Much of the negotiation which was necessary at the time of change to ENG concerned film and videotape editors, whose skills and status were reflected in the different ways recruitment and training were structured. The outcome in many news organizations was the creation of a single picture editor category which recognized the need for the practitioners to be able to handle the entire range of material in any format. All this helped to increase the flow of expertise in unexpected directions, and those writers with a developed sense of things visual no longer feel surprised or upset when an enthusiastic picture editor with a feeling for words suggests a possible line for the commentary.

So, in many ways, it has blossomed into a genuine, two-way relationship, in which more is expected of the editor than the slavish matching of tape to editorial orders. And it remains significant that in the best-ordered news services the commentary is planned around the pictures, and not vice versa. None of these valuable relationships can exist where journalists edit their own pictures or where news is merely one 'customer' for picture editors handling all types of programme material. In these cases editorial staff may be required to view the rushes on domestic type video recorders, noting the length and placing of each shot to construct a cutting order for the editor to follow at a later stage.

Where the two skills are practised alongside each other, picture editors are able to understand how their main problems are usually identical to those of their editorial counterparts – lack of programme space in which to tell the story and lack of time in which to meet an approaching deadline.

In these days of longer newscasts, an editor is probably asked to assemble reporter-made packages much more often than the bread-and-butter items which writers used to script regularly as an everyday part of their work. But the principles have not changed, and any editor aspiring to more challenging work first has to master the techniques on which editing is based, because it really is remarkable how much can be told in 30 seconds of screen time.

This represents no more than six or seven shots, yet if they are put together skilfully the item will make just as much sense visually if dug out of the archives next year as when viewed as part of a newscast in an hour's time.

Sometimes the choice of material to edit may be so limited as to make the picture editor's task one of simple assembly. On other occasions he or she will be overwhelmed. Much depends on the camera-operator who, given a reasonable amount of time on location, aims to provide a series of shots for selection without falling into the temptation of recording everything in sight just because it costs no more to fill up an entire cassette, which is reusable anyway.

Editing usually takes place in any one of a number of special cubicles, the nearer the newsroom and transmission point the better. But although there is no such thing as a 'typical' editing suite, there are a few common features, because whatever system is in use the fundamentals of editing remain the same: the raw material from the camera – the master tape – is never actually 'cut'.

News film was viewed and then broken out by hand into individual shots which were joined together in sequence with clear sticky tape or liquid cement. Sound tracks were edited separately in the same way. Electronic images and sound from videotape are re-recorded – to the lay person with scarcely noticeable loss of quality – on to fresh tape, leaving the original intact.

To do this, every editing area needs to be equipped with two linked video cassette players and monitors (television screens), one to display the rushes, the other on which to build up the edited story. Some suites have a third machine, which allows the editor to include mixes, fades and similar effects; multi-format areas can cope with more than one video recording system. There will also be loudspeakers and an edit controller, a slim box by which the picture editor builds up the pictures and sound. All this equipment can be housed comfortably on a single workbench.

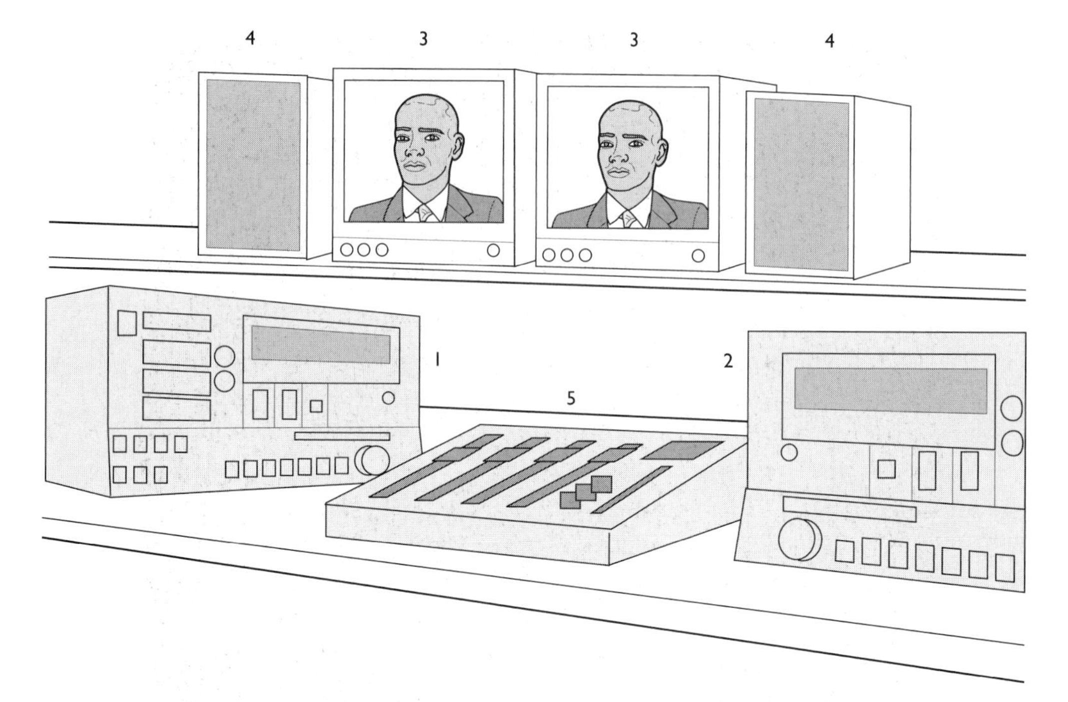

Figure 6.3 Layout of a typical suite for traditional electronic editing. Cassette player (1) for the unedited raw material (rushes); recorder (2) on which edited story is assembled; monitors (3) for viewing the pictures; loudspeakers (4) for the sound; edit controller (5) for assembling pictures and sound tracks.

Although the journalist may not be present during the entire editing process, he or she will invariably try to be there when the picture editor starts work. The story may be spread over several cassettes for which there may or may not be accompanying paperwork to help with identification.

The first step is always to view the pictures. The editor will make sure the power is on before inserting a cassette (tape first) into the horizontal slot in the front of the player. Within a few seconds the tape is automatically threaded and a still picture appears on the monitor. From there the tape is viewed at normal speed, seen 'shuttled' up to ten times faster, 'jogged' gently in either direction using the search dial, or simply fast-forwarded or rewound. A time counter indicates the position of the tape in hours, minutes, seconds and frames. When it is necessary to view another tape the eject button is pressed and the cassette automatically disengages.

The sound from the tape's two tracks comes from a loudspeaker controlled by the sound mixer, which allows the volume to be adjusted to any desired level. As the writer watches, the editor goes through all the picture and sound material, checking for quality and content.

The editor now turns his or her attention to the recording machine, using the 'edit/in', 'edit/out' and 'entry/stop' buttons in much the same way. Selecting 'preview' then allows the edit to be rehearsed. Although the pictures are displayed on both screens, the actual checking is being made on the recorder only. If the picture editor is unhappy with the rehearsal it is possible to go through the whole procedure again, selecting new editing points. If fine adjustment by a few frames is needed the 'trim' buttons are brought into play.

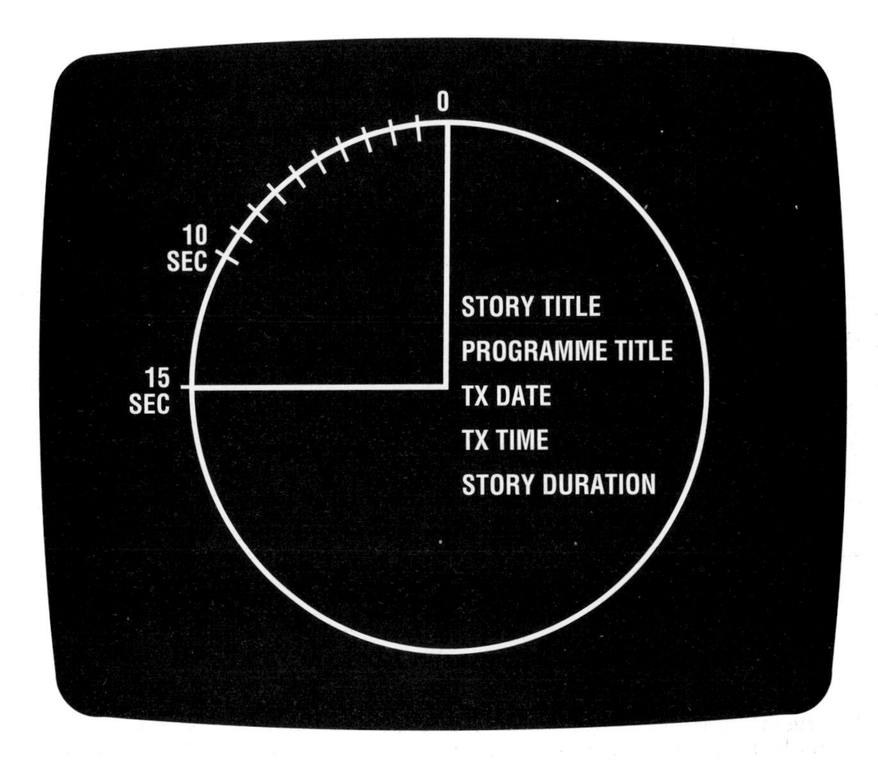

Figure 6.4 Although they vary in shape and form a clock or 'leader' is at the top of every videotape report to ensure it is inserted accurately into a programme.

If there is merely uncertainty that the edit is correct, 'preview' can be used to rehearse the scene once more. But if the editor is completely satisfied, the shot can be recorded from the rushes tape to the new tape in what is sometimes known as 'electronic splicing', by pressing 'auto edit'. When this mode is engaged both tapes roll back three, five, seven or ten seconds from the start of the shot and automatically move forwards again to make the edit.

The first scene should now be safely recorded, but it is possible to check it on the monitor by means of the 'review' button. Once that is accomplished, the picture editor goes on to find the second chosen shot from the rushes, lines up the end of the first already on the recorder and plays it across in the same way as before. The process is then repeated each time with succeeding shots until the whole story is successfully assembled. At any stage the editor may replace a shot or add sound (possibly the commentary).

Long and complicated news features, especially those incorporating graphics or several sound edits, may take several hours to put together. The journalist will probably leave the picture editor and return to the newsroom for other chores. At some stage the writer may be called back to give an opinion as editing progresses, but, as often as not, there is no more contact between the two until the editor's work is completed. Then, after running the story through, a decision is taken at once about whether the cut version comes up to expectations or whether changes are necessary.

Figure 6.5 A tense edit in progress. The edit process is how the 'grammar' of television news is applied. The original material is separated into sound and vision and reassembled so that time is compressed to provide the viewer with the story in a neat complete package. In news, editing is usually against a deadline. (Photo courtesy of and © Sky Television.)

Shot-listing and the script

The writer must also take what may be regarded as the most critical step in the entire operation to ensure that the written commentary matches the edited pictures. Reporters usually lay down their commentary as the editing takes place, one sequence at a time, or, if they are very confident about the material, record the entire report and work with the editor to make the words and pictures fit.

It depends on the nature of the news report. It really mostly depends on whether the journalist knows exactly what sound and vision is available. In all cases, to make good television news the journalist cannot write a single word without knowing exactly what the viewer will see or hear at a given point. For good visual material the best way to ensure that the journalist's words work well with the sound and vision is to do what's known as shot-listing. This consists of noting details of the length, picture and sound content of every separate scene in a sequence. How this shot-list is used depends on many factors: the length and complexity of the edited item, or whether there is time to edit the pictures and then record the commentary before transmission deadline.

Shot-listing procedure is simplicity itself, however long the edited tape, although for this example we shall take an imaginary, typical 30-second story about the arrival of a party leader for a political conference.

The picture editor sets the counter on the cassette player to zero. At the end of the first edit the machine is stopped so the writer can put down on paper everything the scene contains, together with the clock time at the end of the first shot, say three seconds:

GV [general view] exterior of conference hall	3 sec

The machine is restarted and the pictures allowed to run on until the next shot, which lasts four seconds. The writer makes a note of the details and the cumulative time:

MS [medium shot] delegates arriving on foot	7 sec

This operation is repeated until the end of the edited story and the writer's shot-list looks like this:

GV exterior of conference hall	3 sec
MS delegates arriving on foot	7 sec
CU [close-up] crowd waiting	10 sec
LS [long shot] leader's car turns corner	15 sec
MS motorcycle escort dismounts	18 sec
CU car door opens, leader gets out	24 sec
GV leader walks up steps into building	30 sec

Armed with these details back in the newsroom, the writer will be able to time a reference to the party leader to the precise moment, 18 seconds from the start, when the car door opens and that familiar figure appears. Without that information to hand, accurate scripting would be impossible.

Non-linear editing

In many ways non-linear editing (NLE) marks a return to a flexibility which disappeared with the introduction of video. Many years ago one of the benefits of film was that editors were able to join picture sequences into the desired order with sticky tape, and it was equally simple to rearrange them after undoing the joins. Because of the technology, video sequences could only be put together sequentially and any subsequent changes made by repeating the process of transferring shots from one tape to another.

Computer-based non-linear editing machines store the original material on to hard or optical disks, from which shots can be chosen in whatever order is required. When the selection is replayed as an 'assembly' the computer picks out the shots from wherever they happen to have been saved on the disk. Any rearranging, deleting, adding and trimming is carried out without the need to reassemble all the footage, as with tape-editing methods.

The product of a non-linear editing session is the edit decision list (EDL), which is then used to re-create the item from master tapes in a process known as conforming.

Other advantages of the modern system include a compatibility with DV camera and transmission equipment, reliability, and economical maintenance costs compared with broadcast-quality video recorders, whose replay heads wear out more quickly.

Figure 6.6 Digital and disk-based editing. For picture editors or operational editing journalists it brings the flexibility to assemble shots non-sequentially and to make changes without reassembling the entire item. Edits made directly onto high-speed disk drives are available for instant playback. Avid Technology's 'NewsCutter' has two video monitors, a user interface monitor displaying menu bar and editing windows, and a full-screen playback facility for displaying footage. (Photo courtesy of Avid Technology.)

The principal use of machines made by the leading manufacturers has been in the 'off-line' state. The cost of these machines has become economical enough for direct on-line use by news programmes as a matter of course, without the need to go through the conforming process, so saving hours of post-production work. Although the raw videotape rushes are transferred to the non-linear editor in real time, the speed of the editing process itself is seen as unbeatable in its ability to produce different versions of the same story in seconds or to vary its length according to individual programme needs.

7

Writing to pictures

This had been the big weapon in every television journalist's armoury for many years. No matter how much style you might have on screen, no matter how many ideas you have at the editorial meeting, no matter how fast and inventive you are in navigating the web, if you cannot write, you'll be in trouble. The mechanics of writing any sort of commentary to pictures can be *explained* and understood in about an hour. To *apply* them well requires a special ability to appreciate the value that moving images have in relation to the words necessary to complement them. Writing to news pictures is a distillation of that special skill which some journalists, despite a genuine feeling for words and an empathy with pictures, never quite develop.

Basic commentary construction

Exactly what distinguishes the excellent from the merely acceptable is virtually impossible to explain without the impact of the pictures themselves. Looking at the written script by itself will provide no clue. The purists would shudder at the use of the two-word, verbless, inverted sentences and the apparently casual regard for punctuation. The test is to ignore the script and to sit back, listen and watch as the commentary adds a delicate counterpoint to the pictures.

Probably the first mistake the novice writer makes is to try to cram into every second of the available screen time the maximum number of facts previous journalistic experience has taught as being essential. The result is chaos. The words take little or no notice of the pictures they were meant to accompany; the style is heavy, as written for the printed page and, most likely, the reader will come a poor second in the oral sprint to finish the commentary before the tape runs out.

At three words a second, a sequence of pictures lasting, for example, 30 seconds, gives the writer a maximum of 90 words to play with. No matter how cleverly they are used, there is no way in which it is possible to squeeze in more and still expect the commentary to make sense to those hearing it. From the outset, the writer must learn to exercise a ruthless economy of words, first so that the pictures are able to do their work properly and, second, to avoid the ultimate sin of having them finish while the reader is still speaking. It is far safer to under-write and leave a few seconds of pictures unscripted.

Most beginners' commentaries tend to refer in great detail to people, places and events which do not appear at all. This, in many ways, is an understandable fault, but one which must be corrected at once. Over the years viewers have come to recognize such references as signposts leading to what they are about to see, and they are bound to feel cheated if, in the end, those signs lead nowhere.

Too much detail has, equally, the effect of drawing attention to what may be missing from news coverage. A blow-by-blow account of cars screaming to a halt, armed men tumbling out and shots being fired during a jewel robbery should be avoided when all the camera is able to record in the aftermath is a solitary police constable walking over broken glass from a window, a few specks of blood on the floor and tyre marks on the road. The atmosphere can be conveyed just as effectively without using words which make the viewer feel let down that the action is not taking place on the screen.

Similarly with sound: 'cheering' crowds, 'screaming' jet engines, the 'crackle' of small-arms fire – all conjure up definite mental pictures. If the viewer does not hear what is generally accepted as a cheer, scream or crackle, the suspicion may be born that the television people do not really know what they are about. Exactly the same response will be evoked by talk of the 'booming' of artillery, when what can be heard quite clearly on the sound track is indeed the crackle of small-arms fire. In such cases the writer is well advised to use general words less capable of misinterpretation. After all, 'gunfire' is a term capable of being applied to virtually anything between a few pistol shots and a full-scale battle.

Having learned these early lessons, the new writer's next mistake will be to write a commentary which reads like a series of newspaper captions. With every change of shot the viewer is treated to nothing more or less than a verbal repetition of the sights and sounds unfolding on the television screen a few feet away. Thus the writer's influence is as good as meaningless, especially where the script includes phrases intended to ensure that the viewer does not escape even the most obvious:

> 'As you can see here ...'

> 'The Prime Minister, on the right ...'

True, there are occasions when it is necessary to take the viewer metaphorically by the scruff of the neck:

> 'The Smiths knew nothing of the explosion until they returned home a week later. Then, all they found ... was this.'

But it is a device to be used sparingly.

In most cases, to repeat exactly what is happening on the screen is to waste a great opportunity to tell the viewer something worthwhile. The writer's skill lies in being able to convey what is not clear from the pictures.

Take almost any international conference. Ten minutes after the routine photocall, during which delegates are seen talking and joking and filling glasses with mineral water, a furious row breaks out in the privacy of the closed session. Probably all the writer will ever have to work with are pictures suggesting that all was sweetness and light. Instead of throwing away the apparently irrelevant, the writer should be able to make almost a virtue of the scenes of accord, using them to point up the contrast between events occurring before and during the conference:

> '... but the spirit of cooperation didn't last long. Almost as soon as the conference started ...'

The use of archive material when no other illustration is available poses a similar test of ingenuity. Then the writer may be faced with the daunting prospect of matching the apparently unmatchable – out-of-date pictures with up-to-date facts. The temptation here is probably to 'talk against picture', in other words to ignore what the tape shows in order to get the story across. This can be done for very brief periods in a commentary, but the technique needs careful judgement to ensure that words and pictures meet often enough to avoid confusion.

Here's an example. News breaks that the civilian government of an important South American country has been overthrown in a military coup. No pictures are expected until later in the day, and the only material immediately available in the archives consists of a few (unused) shots of the President reviewing troops when he took office on Independence Day a few months earlier.

Armed with the latest information provided by the wire services, a few background cuttings and a carefully made shot-list, the writer should be able to give the viewer a credible preliminary idea of the current state of events. In about 30 seconds the story is told simply enough, even though the pictures and the words accompanying them do not have to be anything special.

Imagine that some of the details of the coup, together with a map and the first political reactions at home have already been given in the studio introduction: the words 'library pictures' will be superimposed on the screen so the viewer is not misled:

Shotlist	*Cumulative time*	*Commentary*
Long shot troops marching through city centre	0 sec	It's not even a year since army generals pledged their loyalty to the civilian government.
Medium shot infantry passing	6 sec	Then they promised to stay out of politics, whatever the result of the elections. But in the capital
Medium shot tank column	12 sec	today the tanks were out to enforce a curfew and people were ordered to watch
Long shot crowds waving	17 sec	state television and await official news. Most of the
General view government ministers on saluting dais	20 sec	leading politicians are now under house arrest, although what's happened to the
Close-up President on platform	24 sec	President isn't clear. He's reported to have had warning of the coup and taken
General view reverse angle	29 sec	his family out of the country at
	31 sec	the weekend.

One of the best tips about scripting good pictures is don't. The more the action, the greater the need to say less. The same principle applies to good sound: let the band play, the cheers ring out. When words are needed in quantity it is important to use them to their best advantage. Too many experienced writers use up all their most interesting facts to cover the early shots and leave themselves short of anything else to say at the finish. Even the commentary to cover a routine 30-second item of local news can be structured to ensure a proper opening, middle and end, instead of being allowed to dribble weakly to a close.

When it is necessary to convey general points, do so over non-specific shots. Learn to recognize how pictures can be made to do their work for you by choosing the most appropriate words to cover them: a reference to millions of pounds being cut from the army equipment budget sits far more happily over a wide shot showing a column of tanks than over a close-up of a soldier on guard duty.

Building in the pauses

As has already been made clear, compiling a shot-list is the only sure method by which the writer is able to identify with any accuracy the separate scenes making up an edited news item.

Applying the formula of three words a second, it would take 54 words to reach the beginning of the shot where the party leader emerges from his car in the earlier example. But this does not allow for the fact there may be some good 'natural' sound to be heard, the writer may not wish to cover all 18 seconds with commentary or, indeed, there may not be enough to say that is relevant or worthwhile.

Added to these factors is the possibility that a hesitation or 'fluff' may make the reader's speed vary, so by the time those 18 seconds have elapsed, words may be significantly out of step with pictures. What is needed, therefore, is some measure of control over the reader while the item is being transmitted live.

The example which follows is a variation on The Party Leader Arrives. This time it is The Foreign Minister of A Friendly Country.

The minister is about to conclude a big trade deal which includes the sale of military equipment for defence against an unspecified external threat. Others fear it may be used to quell internal unrest. Outside the building where the signing ceremony is to take place, police struggle to hold back a group of demonstrators who are waving banners and shouting loudly 'Food not guns!' As he arrives, the minister ignores the crowd and goes straight into his meeting.

Shot-list

General view minister's car and escort	3 sec
Medium shot police link arms to restrain crowd	7 sec
Close-up minister out of car, waving	11 sec
Medium shot group with banners shouting 'Food not guns!'	16 sec
Minister walks straight past, up stairs into building	22 sec

Consulting this shot-list back in the newsroom, the writer decides the most interesting part of the story comes in the third shot, between 11 and 16 seconds, when the demonstrators make their presence felt. To emphasize the point it is essential the chants of 'Food not guns!' should be heard without the accompanying distraction of the reader's voice. The aim, then, is to hold the reader back for the five seconds of the chanting, and then to give a signal to restart the commentary immediately afterwards.

John In Vision

The trade deal between Britain and a Friendly Country has been signed despite controversy over the sale of military equipment. The ceremony, which took place in London, was attended by the Foreign Minister of a Friendly Country in the absence of the President, who is unwell.

VT Insert and OOV (22")
As the minister arrived, large numbers of police were kept busy holding back groups of demonstrators determined to make it known what they thought of the arms deal.

Upsound at 11" (chanting)
Cue at 15"

But even if he heard them, the Minister didn't appear to notice the protests.

Dur 27" (15" + 22" VT)

Figure 7.1 Building in the pauses in the 'Arms Deal' script. OOV is news shorthand for Out Of Vision. This means the presenter keeps talking but his or her face is out of vision and the screen is filled with sound and vision. It would take a professional reader 10 seconds from the beginning of the video to cover the shots of the minister arriving and getting out of his car. The word 'cue' is the sign for the reader to pause while the chant 'Food not guns' is heard. Once 15 seconds have elapsed the reader is signalled to resume the commentary.

The secret lies in part in the format of the written script. If handwritten it might be best if laid out with technical and production instructions on the left-hand side and text on the right. If written directly into the electronic computer system technical instructions may be in a different colour and font, usually red.

Often it is the writer who 'cues' himself to start talking:

> 'As the Minister arrived, large numbers of police were kept busy holding back groups of left-wing demonstrators determined to make it known what they thought of the arms deal.'

In this case the reader will have reached the words '... thought of the arms deal' in exactly 10 seconds. He or she pauses. At 11 seconds the point is reached where demonstrators begin chanting. The shot has changed during the one second it takes for the reader to respond and smoothly pick up the thread of the commentary, so by the time the words

> 'but even if he heard them, the Minister didn't appear to notice the protests'

are heard the VIP is starting to walk up the stairs. (In planning the script it is essential for the writer always to remember a nominal one second between the change of shot or action and resumption of commentary. Experience will tell whether that is a long enough gap.) It is generally recognized that only the writer knows intimately what the pictures contain and is in a position to make minor adjustments to the time to speak.

Getting started

Writers vary in the way they set out to construct their scripts. The method favoured by many is to begin by writing the words around one key shot, not necessarily at the start of the story, and then building up the commentary before and after it, fitting in the cues as necessary.

At this stage, unless pressure of time is great, it is wise to put down more or less anything which sounds right, leaving any polishing of words and phrases until the first draft is complete. This method is further helped by writing only three words on a line, whether on paper or on the screen, so making it simple to add up the number of seconds of commentary already written. The words can be typed three to a line on a blank or customized page, and pasted into the computer system script format later.

However it is done, the rule remains the same – accurate storytelling in news can be achieved consistently only as a result of writing when the shots to be used are known at the *time of writing*. It is disappointing to discover newsrooms where no demands at all are made on writers to 'hit the shots' as a matter of routine. Pictures are allowed to run their course while the words simply wash over them, meeting (if they ever do) more by luck than judgement. What the poor viewer is meant to make of it all is hard to tell.

Finally, if things do go awry, far better the script anticipates what is about to happen than lags behind what has already taken place. The viewer ought never to be left in limbo, staring at a brand new shot, and wondering whether he or she is not hearing the commentary because something has gone wrong with the television set.

Scripting sports news

It really is quite surprising how many experienced television journalists who have no difficulty in explaining the most complicated or abstract issues are completely lost when scripting news about sport. But it is a fairly safe bet to say that large numbers of people in the audience probably know more about sport than they do any other subject. Get the name of a politician wrong on air and for sure some people will be bound to notice: mispronounce some sporting celebrity and the telephones won't stop ringing for days.

What we are considering here has nothing to do with the live broadcasts or other sports programmes which are usually the preserve of specialist departments, but the brief edits for inclusion as part of routine news programmes, as there are times when sport is news and deserves to be treated as such.

Every sport is a subject in its own right, but unless a television news service enjoys the luxury of having its own resident experts, there are bound to be occasions when the writer who is happiest dealing with international crises is suddenly called upon to script a tennis final or a soccer match. Sport has a language of its own, with accepted terms and phrases recognized by the aficionados but which are inapplicable, inexplicable even, out of their own context.

Simple things: teams take the collective noun. It may be more grammatically correct to say 'England is ...' but anyone who is the least bit familiar with sports coverage of team games will pour scorn on any organization which allows that to be broadcast in place of the accepted 'England are ...'.

The second, and much more difficult problem, concerns the construction of the script itself, because the action in most sports is telescoped into a very short space of time. Long before any accompanying script can explain what has happened, the golfer has sunk a winning putt, the tennis player has served an ace, the batsman has been clean-bowled. So the most sensible approach to scripting virtually any sport is to remember that the action is far better 'set up' before it takes place, with any additional information immediately a suitable pause occurs – as the golfer picks the ball out of the hole, the tennis player winds up for the next service, the batsman starts to walk back to the pavilion.

The edited opening 13 seconds of a report on an England soccer match should give an idea. The script is simple, straightforward and economical. Note how the shot-list splits the action into very short scenes to ensure maximum accuracy in the commentary writing:

Shot list	Cumulative duration	Commentary
GV interior stadium and spectators	0 sec	A capacity crowd of eighty thousand saw England quickly into their stride,
MS Owen, no.9 is tripped going towards goal	5 sec	Owen, at number nine, soon proving a handful for the defence. Beckham's
WS Beckham takes free-kick	9 sec	free kick – perfection:
MS Owen leaps	10 sec	
MS goalkeeper is beaten	11 sec	(Cue) Revenge! Owen after six minutes.
CU Owen starts to celebrate	13 sec	

Note how the opening words set the scene and at the same time warn the audience that something important is about to happen. They also straddle the second shot, identifying the player who has been tackled illegally, and there is still enough time to squeeze in the information about the taker of the free-kick before the commentary pauses at nine seconds to let the action take place.

During the next four seconds the free-kick is taken, Owen leaps into the air and heads the ball past the goalkeeper. It is only after that, at 11 seconds, as the scorer turns to celebrate, that the commentary is allowed to resume, completing the sequence while confirming in words what the pictures have shown.

It is a treatment which probably works equally as efficiently with other sports, the essential point being that the script always helps to build up the expectation of important action and does not swamp it with unnecessary words.

Finally, it is also true that some good journalists who do know about sport somehow go over all glassy-eyed when they are writing about it. Because it is a subject they love and know intimately, they allow themselves to relax, and one result is that many of the safety devices they have invented for themselves get ignored, standards drop – and suddenly more clichés. More tame and predictable phrasing appears in sports writing than in any other part of the newscast. It is essential for the enthusiasts to remind themselves they are writing for a general audience who have every right to expect sports news to be treated as skilfully as any other subject.

Cueing into speech

Some of the problems associated with cueing natural sound effects within a news story have already been touched on. A whole new set of difficulties present themselves when the sound is that of human speech.

Much of what is spoken within television news reports comes from two categories of people – reporters or those being interviewed by them. In either case, the aim of the writer must be to construct any additional commentary for transmission in a way which links most naturally into the words already recorded on to the sound track.

Where this comes within the body of the item rather than at the beginning, the onus rests even more heavily on the editorial team to ensure that speech arrives a decent breath's pause after the commentary introducing it. Failure results in either an embarrassingly long delay between the two or, worse, what is known in the United States as 'upcutting', the ugly overlap of live commentary and recorded sound.

Accuracy is achieved fairly easily by positioning a cue paragraph immediately before the sound extract. The wording itself is of importance, and the writer is taking unnecessary risks if a planned lead-in to a section of speech demands timing to the split second:

> 'Turning to the latest round of trade talks, the Foreign Minister told the conference ...'
> *(Foreign Minister speaks)*

will be impressive if the sound is heard without delay, but

> 'The Foreign Minister told the conference about the latest round of trade talks ...'
> *(Foreign Minister speaks)*

would be much wiser, for it would still make sense if the recorded speech were delayed for a few seconds or did not arrive at all because of some technical fault. The whole principle is based on the fact that flexibility rests with the writer's words and not on the speech fixed at an immovable point on the edited sound track.

With interviews in which the reporter's first or only question has been edited out, the words leading up to the answer must be carefully phrased to produce a response which matches and makes sense, otherwise there is a clear danger that the writer will be guilty of distortion. It is equally important to ensure that if the viewer is about to be shown an interviewee in close-up preparing to answer a question, the commentary leading up to it should leave no doubt about who is to speak. This is achieved by referring to the first speaker last. So it is:

> 'Tania Bailey asked the Foreign Minister for his reaction ...'
> *(Foreign Minister answers)*

and not

> 'The Foreign Minister talked to Tania Bailey ...'
> *(Foreign Minister answers)*

All introductions should follow similar principles.

> 'We've just received this report from Delia Ward ...'

as the final sentence leading into a story about farm subsidies is acceptable only if Our Reporter is visible or audible at once. But if the opening shot shows an impressive herd of prize-winning cows Ms Ward is unlikely to be best pleased by the inevitable audience reaction. A less general scene, perhaps of a field or farm entrance, would reduce the disparity.

It is also worth bearing in mind the use to which final introductory sentences can be put as a way of alerting the viewer to what is coming next. The flat statement:

> 'Delia Ward reports ...'

is factual but adds nothing.

> 'Delia Ward has been finding out why the withdrawal of farm subsidies could have such devastating effects ...'

has, at least, the merit of suggesting to the audience that it might be worth keeping awake for the report about to follow.

Selecting soundbites

The selection of one or more extracts from a lengthy recorded interview obviously depends on the amount of space the item has been allocated within a newscast. As it is fairly unusual for any interview to be shot to its exactly prescribed duration, however experienced the principals taking part, a certain amount of choice will inevitably be necessary. Given time, the programme editor or producer may wish to make this part of their normal duties, especially when sensitive issues are at stake, but just as frequently it is the newsroom-based writer who will be faced with the task, sometimes 'guided' by the interviewer to the most appropriate quotes. In some news organizations the reporter is expected to 'cut' his or her own interviews without any supervision.

But as every interview is unique, it is impossible to set down rigid rules. On some occasions, the single one-minute answer out of six will stand out. On others it will be difficult to quarry a twenty-second bite from fluff-ridden ramblings.

The firmest general guidelines to any selector probably go no further than a suggestion that the ultimate value of any news interview rests more in 'colour', opinion and interpretation than in the presentation of facts, especially those which may become quickly out of date or easily challenged. What the audience wants to know from officials interviewed at the scene of a fatal accident an hour before the newscast is how it might have happened and not the number of deaths and injuries – which could be liable to change. That information is best left for the studio introduction or updated at the last moment for inclusion within the overall report.

In making the choice, there should also be awareness that much more is involved than suitability of duration and content, important though they might be. The editing into, or out of, any recorded speech at precisely the required point editorially may, at the same time, not be feasible technically. So even at the expense of a few, extra, unwanted seconds of screen time, the aim should always be to cut at the most natural points: ends of answers, or where the selection consists of only part of a sound passage, a stop or breath pause during which the inflection of the voice is downwards.

Although most people are fully aware that editing takes place, it is always much better to avoid any cut which will appear both ugly and obvious.

Last words about pictures

Two temptations to avoid are puns and clichés.

Experienced writers usually consider a really important, well-shot news story virtually tells itself, the task becoming one of assembling facts in an order dictated by the quality and sequence of the accompanying pictures. Much more testing are the down-bulletin items, often weekend fillers or 'soft' stories for which little information is readily available. With these, the temptation is for the writer to produce a stream of generalities or a series of puns, the aim in either case being to lower a curtain of words through which the lack of facts will not be noticed. There are occasions when, used sparingly, this technique does work. But, for example, a balloon race which wrings out such lines as 'soaring reputations' or 'rising hopes' will quickly have the discerning viewer zapping channels in disgust.

For any unwary writer, the cliché presents another booby trap, and in television news it is a double-edged one at that, since trite pictures are just as likely to find their way on air as are trite phrases. Probably every viewer of every television newscast in the world has had to suffer the local equivalent of the following British examples, from which not even the most lauded, high-minded news programmes are immune:

- exteriors of courthouses or other public buildings, used to avoid reporter pieces to camera;
- any politician sitting at a desk signing documents or reading letters;
- Cabinet ministers filing through/coming out of the door of No. 10 Downing Street;
- any VIP descending/ascending any aircraft steps anywhere;
- camera crews or security men on rooftops, used to telescope the action between aircraft steps and official car;
- crowd 'reaction'.

As for the words, it seems almost impossible for some writers to avoid trotting out the stock phrase to satisfy the stock situation:

- The Big Fire
- 'Fifty/a hundred fire-fighters fought/battled the blaze/flames'
  'Smoke could be seen five/ten/fifty miles/kilometres away'
  'Fire-fighters/ambulances rushed ...' (what else would they do?)
- The Explosion/Earthquake
  'Wreckage was scattered over a wide area'
  'Rescuers tore at the wreckage with their bare hands'
  'Damage is estimated at ...'
- The Injured
  '... undergoing emergency treatment' (see 'Fire-fighters/ambulances rushed')
- The Great Escape
  'Police with tracker dogs ...'
  'A massive hunt/search ...'
  'Road blocks have been set up'
- The VIP Visit
  'Security was strict/tight'
- The Long Drawn-Out Negotiation
  'The lights are burning late tonight ...'
- The Holiday Snarl-up
  'Traffic was bumper-to-bumper'

- The Appeal
  'Unless the Government provides more money/changes its mind/a donor is found/comes forward, the department/hospital/project/child will close/have to move/die.'

And that is not to forget my own particular favourite, usually attributed to an elderly eyewitness to any violent incident in the south of England:

'It was just like the blitz.'

To be fair, it is perfectly understandable that when time is short and the pressure great, it is the familiar line rather than the elegant phrase which suggests itself to the writer, besides which the overriding priority must always be to get the commentary on air, however much it might lack in originality.

But that ought to be reserved for the last resort. Where second thoughts are possible, the tired old standby must be shunned. As experienced journalists like to put it: 'Avoid clichés ... like the plague.'

The golden rules of writing to pictures

Writing to pictures presents journalists with a genuine opportunity to extend their experience into a completely new area. Yet, paradoxically, it remains one with limits which some regard as too restrictive. In accepting the first principle that there can be no scripted words totally unrelated to the pictures accompanying them, the writer may feel a straitjacket is being fashioned from the very material it was believed would usher the viewer to new heights of understanding.

Among some professional writers, and especially those making the transition from print or radio without adequate training, this feeling is sincere, the gap seemingly unbridgeable. It need not be, provided that what is an apparent weakness in the whole foundation of television news is seen as a means of refining news sense to a point where every single word is carefully chosen before being put to work.

As confidence improves, the dedicated convert to television news discovers that within the boundaries of content and duration the treatment of words and pictures as complementary in character makes it possible to convey deeper understanding of both.

To begin moving towards that goal, the writer must take time and care to apply, ultimately by instinct, what can only be described in summary as the golden rules of writing to pictures:

1. Words and pictures must go together. Fight the pictures and you will lose.
2. Don't repeat in detail what the viewer is able to see and hear for him or herself. This is television, not radio.
3. Don't describe in detail what the viewer is not able to see and hear for him or herself.
   The audience will feel cheated otherwise.
4. Don't overwrite. The best script is often the one with the fewest words.

And, to repeat:

Match your words to the pictures, not vice versa.

8

Television news reporting

Despite the enormous satisfaction it is quite possible to derive from the business of putting together complicated news stories for transmission in a very short time, and the knowledge that real editorial power lies ultimately in the hands of deskbound executives, there is not much doubt that the glamour image of the average television newsroom journalist lags far behind that of those who appear in front of the camera. For while no viewer would be expected to name any member of backroom teams, descriptions of nationally known news performers – the 'talent' as they are quaintly called in the United States – trip easily off the tongue. So it is hardly surprisingly to find a high proportion of would-be recruits and starry-eyed newcomers to television news hankering after what they believe to be the ideal – to be seen by an audience of millions through news reports made in some exotic, mildly troubled spot on the other side of the world, enjoying what one member of the international reporting set has summed up as 'a front seat on history'. In short, however unsuitable they might be as performers, they want to be famous.

This desire may be categorized as a perfect example of the greener-grass syndrome, for there are certainly some reporters who would dearly love to exchange their front seats for what they regard as the calmer back rows in the newsroom, pulling the strings. Even experienced television reporters privately admit that, after a while, the apparent glamour and excitement of their lives begins to pall. One described it a taxi-rank kind of job. Some learn sooner than others to detest rushing to catch planes and deadlines, living out of suitcases, eating hurried meals in unhygienic places abroad, witnessing at first hand unspeakable horrors of which the audience may ultimately see very little.

All this is in addition to the real personal dangers involved in covering the stuff of modern television news: war, natural disaster, civil unrest. That professional newspeople doing their job are just as much at risk as the combatants has been proved time and again over the years by events in South-East Asia, the Middle East and Central America. The worst recent example was Indonesia, which was considered to be the most dangerous war zone for the press since Bosnia in the 1990s. At times journalists are deliberately targeted. Even the most ragtag of irregulars, for all their apparent naivety, lack of discipline and modern military equipment, seem to be very well-informed about satellite communications and the power of television to influence their cause for good or ill, and behave accordingly.

Worried news organizations now issue comprehensive guidelines on how to avoid problems. Journalists preparing to report from potential trouble spots are checked for their health, sent on survival training courses run by the military, and are issued with flak jackets and other protective equipment before they get anywhere near the front line. Once in war zones they are expected to share the physical discomfort undergone by the troops whose campaign they are covering, keep their heads down when the shooting starts and still meet their deadlines. Being female brings with it no guarantee of special dispensation.[1] Counselling may be necessary at the end of particularly unpleasant or traumatic assignments.

It is not only foreign journalists who are at risk. The New York-based Committee to Protect Journalists (CPJ) has warned that the main threat to journalists comes from non-governmental nationalists and religious extremists. When the violence in East Timor was at its height in late 1999 a sign in the lobby of the Hotel Turismo in Dili said it all: 'Kill all journalists.'

On the positive side, the CPJ's view was that more journalists were getting into trouble because there were more journalists than ever before serving more news outlets then ever before. Dozens of countries where there was virtually no press freedom ten years ago now have thriving independent broadcasting outlets.

Many reporters are married with family responsibilities, and live with the uneasy feeling that any birthday party, wedding anniversary or other normal domestic occasion may be interrupted by a sudden telephone call commanding them to be on the next flight to somewhere or other. The wife of one former television news general reporter used to say the one thing which unnerved her above all was the sight of the small suitcase containing spare shirts, underwear and shaving kit, which stood permanently in the hall as a daily reminder of the emergency assignment which might come at any time: that, and not knowing whether the news 'event' would last a couple of days, a month, or more. Understandably, not all relationships endure that sort of strain.

Of course, no sensible reporter pretends it is all hectic. Most will readily recall hours wasted at airports, in draughty corridors of government or other buildings waiting for events to take place or people to turn up. Sometimes they did not. At other times doors slammed, telephones went dead, the answer was 'no comment' or something less polite.

In contrast there are pleasant, well-ordered and interesting assignments during which the reporter enjoys top-class travel, hotels and food, is met with enthusiasm, generous hospitality and a genuine invitation to call again. As a bonus there is membership of a small group of privileged witnesses to history as it happens. 'Reporting television news is the only profession in which you can fly 5000 miles, drive 200 miles to a town you've never visited before, walk round the corner and meet 52 people you know.'[2]

The result of a job well done may be two or three minutes of good pictures, a visual by-line and an enhanced reputation, yet the dominating factor is that – apart from those occasions where circumstances demand otherwise – the reporter cannot be expected to work alone if the assignment is to be carried out properly. For while the solo newspaper journalist and the radio reporter in the field are as close to the office as the mobile telephone in their pocket, the television news reporter has to work with a camera crew, a satellite or microwave links unit or a television studio linked in some way with home base. Even in these days of compact, mobile equipment, this is bound to put the television news reporter at a disadvantage when it comes to the scrimmages which are inseparable from so much world news-gathering. There is little point in pushing through a crowd of equally pushy newspeople, only to discover that the camera-operator has been left behind struggling with his gear.

1. See Anna Sebba (1994). *Battling for News: The Rise of the Woman Reporter*, Hodder and Stoughton.
2. Michael Cole, formerly of BBC News, quoted in *Executive Travel*, September 1985.

Yet to be effective the television reporter cannot be content to hover aloof on the fringe of a story in the hope of eventually being granted special treatment by the other participants. Despite the fact that the obtrusiveness of the camera, microphone and lights makes the team a target for attention and occasional abuse, any reporter who is not up with the herd and sometimes in front of it does not last very long in the job.

And then there is the irony that perseverance and initiative at times work to the reporter's own detriment. How often it seems the fruits of a good television interview, grabbed against all the odds, are picked up by other newspeople in the pack and, with minimal embellishment, are turned into highly acceptable accounts for their own branches of the news business. At other times during the scrum it is impossible to discover who has asked the questions which are eventually heard on the sound track, but that has not always stopped the critics from blaming television reporters for any crassness, poor grammar or apparent insensitivity.

In the end, though, the reporter for television news is only as effective as his or her last report. There is no glory to be won from the production of a brilliant piece of work which arrives too late for transmission. Time, effort and money are wasted if, at the conclusion of an expensive foreign mission, the tapes are confiscated, never to be seen.

Reporting as a career

Few television reporters begin their careers as such. Most graduate from newspapers, news agencies or radio, and so lack only a knowledge of television techniques to become successful. Others are recruited for their specialist rather than journalistic experience. Some in both categories fail completely, while others turn out to be competent without ever fully understanding how to construct the well-turned phrase which complements rather than competes with the pictures, and it is not surprising that the best exponents of the reporting art are often those who have served a period as newsroom-based writers.

Leaving aside the matter of journalistic ability, a modicum of which must be assumed, the two basic qualities every reporter must have before being let loose in front of an unsuspecting world seem to be a reasonably personable appearance and clear diction.

In Britain, ideas have undoubtedly changed about what constitutes diction good enough for broadcasting. The general increase in news outlets, particularly since the arrival of local radio in 1967, has allowed all manner of accents and speech impediments to become suddenly more acceptable. Accents don't matter if the voice is clear and does not contain distractions. However, the cruel truth about television is that even expert journalists can lack credibility in front of the camera in the studio, or on location. The most carefully researched, well-crafted piece of journalism is totally lost to the nine-tenths of an audience fascinated instead by a nervous tic, bobbing Adam's apple, dental problems or an inability to keep the head straight.

That is not the only difficulty. What some viewers consider to be incorrect pronunciation is guaranteed to induce near apoplexy on the part of the critics, amateur and professional alike. Television critics and newspaper columnists regularly reprimand certain broadcasters, television news reporters included, for sometimes pronouncing the word 'the' as 'thee'. It is worse with place names. Get a town in Wales wrong and you'll know all about it! Imagine you live there yourself and you hear it totally mispronounced.

Still, at least such errors are capable of being corrected, unlike the expressionless monotones, glottal stops and nasal whines which apparently defy the best efforts of the voice coaches. Little can be done either, it seems, for those with voices so light and high-pitched as to make virtually no impact for broadcasting. A cruel fact is that it is much harder to repair, through training, a difficult voice than it is to improve a person's appearance.

In contrast, chocolate-box good looks and speech which is too precise are considered equally off-putting, whether found in men or women. Among all except those who mourn the demise of the Hollywood glamour factories, the preference is for people who look and sound as though they lead real lives off screen.

All this merely goes to emphasize how easy it is to be critical. Given the curious chemistry at work in everybody's likes and dislikes, it is interesting to speculate on the fruits of a computer programmed to produce a template of the reporter most likely to win universal appeal. No doubt that has already been done. But without the benefit of that research, it must all be down to intelligent guesswork and a single, old-fashioned word – style.

Learning the ropes

The novice reporter quickly discovers there is no short cut on the tortuous route which may eventually lead to general acceptance as a competent television reporter. There is likely to be very little in the way of formal 'coaching', as the average news service – a few enlightened training providers excepted – expects its newcomers to pick up everything except the basics as they go along. This is euphemistically described as 'on-the-job training', and comes as a complete surprise to reporters who may be expected to master the intricacies of 'filming' from their opening assignment.

Once some initial progress has been made, usually after a few painful lessons on the way, the new reporter may be taken aside by a more senior colleague and told gently about some of the most obvious flaws. Some of these might be avoided in the first place by attention to three factors which add to or detract from any on-screen performance: speech, mannerisms and dress.

Speech

Everyone who appears regularly in front of the camera develops a natural, personal style of delivery and emphasis, and although this individuality is to be encouraged, the aim in every case must be clarity, with delivery at an even pace. It must be neither slow enough to be irritating nor fast enough for the words to run into each other: no audience is able to take in much of what it is told by an excited reporter speaking at full gallop.

As part of a general tendency to group words and phrases in a manner which sounds odd as well as ungrammatical, one of the most common is the addition of non-existent full stops in the middle of sentences. The cure could not be simpler: sentences which are too long should be broken up into shorter ones.

Fluffed lines and hesitations inevitably mar otherwise fluent performances and perhaps lead to loss of confidence. Preparation is essential. Much of the time reporters are reading what they have written, so they should be familiar with the contents of the script well before they come to record or broadcast live. Rehearse as thoroughly as possible (mumbling in a corner, though a poor substitute, is better than nothing). Learn to 'project', with good use of voice inflection and pauses to make the script clearer.

When serious mistakes do occur there is no shame in asking for a second take, for even the most experienced performers expect to trip over their words from time to time.

Practice reading with a tape recorder and be as self-critical as possible when listening to the replay. Where faults seem to persist there is no harm in consulting professional speech therapists, who are able to devise little training routines for the tongue which can only increase the performer's confidence.

Mannerisms

In reporters with easygoing, relaxed personalities, tiny mannerisms may become endearing to the viewer. An occasional frown, raising of the eyebrows or head movement to emphasize a point, probably comes across as genuine involvement in the story at hand. For the rest, stiff, awkward movements, facial contortions and continual passing of tongue between dry lips are among the many tell-tale signs of stage fright. Usually this disappears once confidence comes, although not always. I know of one former reporter who, while completely at ease before the microphone in the radio studio, betrayed his nerves during appearances on television by prefacing almost every sentence with the word 'well', even though he knew it did not appear once in the script he had written.

Nervousness is not shared equally between recorded and live performances. In many ways the studio camera seems to magnify mannerisms which, to the consternation of studio staff, reveal themselves only under the strain of live transmission. Some reporters slouch back in their chairs, others tilt like the Leaning Tower or hunch their bodies so that one shoulder is thrust aggressively towards the camera. Possibly worst of all is the fear which has the reporter sitting literally on the seat edge. The result is a close resemblance to a jockey on horseback, except that to the viewer the rider here seems poised to leap out of the set and land in the front room. At least twitching hands are usually hidden by the camera angle.

To all those who suffer from it, this stage fright (no respecter of persons) can become increasingly confidence-sapping. Practice will make the biggest contribution towards overcoming it, especially if backed up by the close scrutiny of recordings of personal performances. The advice of production staff, given and received in the right spirit, will also help the novice to isolate and then dispose of the main problems which, if left to develop, might lead to permanent bad habits.

Dress

The medium itself imposes some restriction on dress: the sensitive mechanism inside electronic colour cameras seems unable to digest certain striped or checked patterns which set off disturbing visual hiccups known as strobing, and some colours (notably shades of blue) create 'holes' through which studio backgrounds appear. Aside from that, some news-type programmes have recognized how dress contributes as much as set design to the identity they wish to create.

Early morning programmes in particular seem to like the casual look, and some ask anyone who appears, visitors included, to dress accordingly. This fits in with the general requirement for reporters to wear what is in keeping with the programmes they are working for and the stories they are covering. For example, an open-necked bush shirt and denims would be entirely appropriate for covering a desert war, while formal business wear would not, and a studio interview calls for more formal clothes. The

Figure 8.1 Preparing for action. Viewers everywhere have a right to expect professional television journalists to present themselves in ways appropriate to the story they are covering. For the reporter above, this meant a quick grooming in public before facing the camera. The scene was captured by the author on the steps of Sydney Opera House.

important thing for reporters of either sex is to avoid clothes and colours which the majority of viewers would consider inappropriate. For women, there are enough smart, businesslike styles available to preserve femininity without resorting to anything fussy, although almost anything women wear on television seems to be regarded as fair game for criticism and comment from fashion writers and others.

As for grooming, it would be unfair to expect the viewing audience to accept uncombed hair or a two-day beard where the reporters' families would not, except in those situations where such an appearance would be relevant to the story.

Beads, jangling bracelets or long earrings are best avoided, as their movements are inclined to create distraction at the wrong moment, especially if they fall off. Lapel badges, in particular those which just defy identification, are fraught with danger. So is the whole range of 'club' ties. The possibility here is that the viewer might miss all that is being said while concentrating hard to see whether the coloured blob three inches below the knot is of real significance or just a gravy stain.

The reporter's role

However the dictionary may define the role, the public perception of the reporter – as influenced by fiction and experience born of reading, watching, and listening to news – is most probably that of the journalist who goes out and 'gets the story'.

But as with so many other areas of television news, reporter duties vary according to the size and importance of the organization being served. In some cases reporters are regarded as the most important editorial animal, responsible for generating their own stories and then seeing them through all the production stages towards transmission; in others every journalist except the editor/producer is categorized as a reporter, no matter how infrequently they leave base; in a few, the general reporter's scope is much more limited, with daily on-location assignments carried out at the behest of senior editorial staff and planners of news-gathering activities, the final product shaped by newsroom-based writers and producers.

In places, the most noticeable movement has been away from generalists towards correspondents, higher-graded reporters responsible for areas of specialist coverage, the aim being to break away from what has been criticized as television's tendency to react to events rather than originate material of its own. Although usually under the authority of the news editor, correspondents are regularly afforded time away from the daily grind in which to research and prepare their stories from their own sources.

Ready for assignment

Any generalists in the team, freelances included, may expect to be given a certain amount of briefing, even if it is limited to the approximate outline any contribution is expected to follow to enable it to take its place within the rest of the newscast. Where an assignment is foreseen as representing only one segment of wider coverage of a single topic, briefing is much more detailed. Good preparation is vital at any time. Given reasonable warning of the nature of an assignment, a diligent reporter will make a virtual fetish of reading up any available background material. On foreign assignments, this may run to dossiers built up from previous visits and include a diversity of facts ranging in importance from currency exchange rates down to the names and localities of reliable laundries.

Travel arrangements vary. To ensure speed off the mark, those news services able to afford it provide individual reporters with bought or leased cars, complete with two-way radio links, or at least contribute fuel and other running costs towards the reporter's own private transport. Others find it cheaper to run a pool of office cars, perhaps providing chauffeurs to drive them so that reporters can be in and out fast without worrying about finding a parking space.

For those unwilling or unable to match such luxuries, reporters are expected to travel with the camera crews or simply jump in taxis, either paying as they go and recouping the expenditure later, or, as part of official account arrangements with taxi companies, signing the driver's log at the end of each journey. Some news services operate a variation of the pool system, ferrying all operational staff to and from assignments in the same vehicle, but this has its drawbacks. There are apocryphal tales of news teams stranded miles out of town or at headquarters, officially unable to move until an office car became available, while some government building went up in flames at the hands of rioters.

Reporter as manager

The changing nature of news-gathering has inevitably had an effect on relationships in the field. Where at one time the news camera-operator was a member of a technical

FIVE O'CLOCK REPORT

Assignment sheet

ASSIGNMENT No. 0417 DATE: 13.6

SLUG: Topfield

CAMERA: Bryan

REPORTER: Delia Ward

LOCATION
Topfield Farm, Dunton, Oxfordshire

TIME: 1100

CONTACT: John Brown (farmer) PHONE NO: 0606
Alec White (NFU rep)

STORY DETAILS:
John & Marjorie Brown are feeling the effects of Euro competition at their farm in Oxfordshire & may have to sell their prize herd of Jersey cows to stay in business.

COVERAGE REQUIRED:
Farm activity. Interview with John Brown.

ADDITIONAL NOTES:
RV 1045 at M40 service station (South side) (Note: may also be wanted for lunch programme.) Copy rushes for Country File.

Brian Flynn
Duty News Editor

Figure 8.2 Example assignment sheet. A disciplined approach to news-gathering means including contact mobile numbers, rendezvous times and, for some assignments, possible hazards, such as proximity to farm animals.

team with a separate sound recordist, the move towards single-crewing has led to greater interdependence between camera-operator and reporter. What has not changed is the 'managerial' mantle, which continues to fall on the reporter apart from those occasions when a field producer is involved, and which covers overall responsibility for the shape and content of coverage.

In between lies a fascinating, ill-defined area of ground which in non-news location work would be covered by a director. For reasons chiefly of cost and mobility, it is generally accepted that television news teams in the field do not need to be accompanied by a separate director, the role being shared by reporter and camera-operator on the spot. So it is probably here that the greatest scope exists for disagreement.

The ideal working compromise consists of a reporter with good journalistic skills and an eye for pictures sketching an outline to be filled in by a sympathetic, experienced camera crew (of however many). Detailed discussion about the best way of achieving the desired end product is advisable before a single shot is recorded. But, in the final analysis, it must be the camera-operator who decides what is technically possible, depending on numerous factors including the available light and distance from the subject.

Once a general storyline has been agreed, the reporter then has to trust the camera-operator to supply what has been promised. Long arguments about the closeness of a close-up or the speed of a pan only hinder the completion of an assignment, and no professional news camera-operator would tolerate a reporter's demand to look through the camera viewfinder before *every* shot.

Relationships are therefore important, particularly on some dangerous foreign assignments, where the degree of mutual trust could make all the difference, literally, between life and death. Sometimes reporters and crews will build up personal friendships and respect over a series of difficult, successfully completed assignments. Between others, the chemistry will be all wrong and no amount of attempted peace-making will put it right. To team a lazy reporter with a go-getting camera-operator, or vice versa and still expect the screen to reflect only successful results is wishful thinking. Far better to ensure, where possible, that incompatible factions are kept well apart.

Even when the prospects for cooperation are good, there is no sure recipe for success. The reporter must always remember to be considerate and tactful in the treatment of professional colleagues, resisting any attempts by misguided outsiders to create separate categories of 'officer' (reporter) and 'other ranks' (crew). Reporters who allow themselves to be swept off to the executive dining room while the camera-operators make do in the factory canteen deserve the inevitable opprobrium.

Equally, the camera-operator must be patient with a nervous or out-of-sorts reporter. After a long, tiring day with very little to eat or drink, it is often tempting to give the thumbs-up to a reporter's performance known deep down to be flawed, just as the timid reporter, suspicious that something may be wrong, is prepared to accept a personal second best rather than risk offence by encroaching on crew mealtimes.

Getting the story right must come first. As one experienced camera-operator has put it: 'If the reporter fails, I fail.'

In addition to the constant awareness of deadlines, there also has to be recognition of the need to be economical in the use of tape, not so much for the sake of cost as for the reason that the greater the volume of material, the longer the time necessary for viewing and editing, a task not necessarily always made easier by the addition of time-coding to recordings.

With an assignment completed and unedited pictures received at base, the reporter's role becomes blurred. One of the main planks of the intake–output system is that it is the editorial staff back at the office who assume the final responsibility for shaping material to include in the newscast. Although the reporter's guidance may be sought, the theory is that those most closely involved in the creation of items are not necessarily the best placed to make objective judgements about their value. This is apart from the possibility that all manner of developments may have taken place which downgrade the original importance of the assignment.

Yet many reporters, as specialists in their subjects, quite understandably resent being told which are the 'best' bits of their interviews, and in some quarters it would be regarded as unthinkable for anyone to come between reporter and story. Modern video packages especially are so dependent on the reporter's ability to mesh the various pieces together that no producer or newsroom writer, coming cold to a project, would probably be able to understand much of it in any event.

Facilities allowing editing on site and the relay of completed reports have further enhanced the reporter's control over their own material, and it would be against human nature to expect them to miss any opportunity of doing so.

9

Reporting techniques

Although electronic and digital news-gathering has long superseded film as the main means of originating news material, most of the reporting techniques have remained unchanged.

Pieces to camera

Of all the skills needed for television news reporting, the piece to camera is among the most frequently used and as such is probably the most susceptible to changes in fashion. For reasons it is hard to fathom some programmes would rather have the viewer looking at a succession of building exteriors than a reporter's face, and experienced journalists are known to consider they have failed if they have to resort to one, but there is no doubt that the 'standupper' – to give its alternative name – remains a sure and straightforward means of giving the news on location.

It has three advantages. It immediately establishes the reporter's presence on the spot; it is extremely simple to execute, and it is fast enough to be considered a kind of contingency sample, rather like the dust scooped up by the first men on the moon in case they had to return to Earth rather hurriedly.

Chiefly because of its speed and the fact that there may be no other pictures to supplement it, the piece to camera can be designed as a complete report by itself, yet it probably has greater value as one ingredient within a comprehensive news report, being versatile enough to be slotted in at almost any point, not necessarily at the opening or closing stages.

The term 'piece to camera' is self-descriptive, being those words which the reporter speaks aloud while looking directly into the camera lens and, through it, to the viewer. The technique depends on an ability to write spoken language and to remember it word for word when delivering it to the camera. But in some respects what matters more is the choice of location for the operation. For example, there seems little point to be gained in travelling thousands of miles and then pointing the camera at the reporter standing in front of some anonymous brick wall. Unless the brick wall is germane, or there are legal problems such as exist over filming within court precincts, the aim should always be to show reporters actually where they say they are. To say proudly that our reporter is there is one thing: to prove it to the viewer is something else.

That does not mean going to ridiculous lengths to find a background which is visually exciting but irrelevant to the story. It should be enough to place the reporter in a spot

Figure 9.1 All news services have their own ideas about how reporters should be framed on the screen. Central positioning may have the effect of masking some of the picture (a). Putting the reporter to one side improves composition, making her part of the action and not just a superimposition on it (b).

which is appropriate, interesting, but not too distracting. If the welcome which greets the news team is not overwhelming, the piece to camera is capable of being completed within a very few minutes, provided the camera equipment has been tested and is known to be working properly, and the reporter is ready with the words.

Most pieces to camera are recorded as the reporter stands full-face to the lens. But putting the reporter to one side of the frame rather than in the centre ensures that any action in the background is not completely blotted out, and the figure seems to become part of the picture and not a superimposition on it.

Sensible variations are to be welcomed as long as they do not seem to be too contrived. On occasions these may be forced upon the reporter by the situation: sitting in aircraft, cars or trains, crouching under fire or walking along a road. So much depends on the styles programmes establish for their reporters to follow.

Knowing the words

Much of the apprehension felt by novices about their first pieces to camera is caused by doubts about their ability to remember their words. Admittedly this can pose a real problem, for it is a knack achieved more easily by some than others.

Newcomers to television news reporting are haunted by the possibility that even a short, apparently simple piece will require several attempts, resulting in a waste of time, temper and a humiliating starring role on the private tape of bloomers television technicians love to compile for showing at office Christmas parties. They can take comfort from the admission by some experienced practitioners that, even under the least demanding circumstances, they are unable to remember more than a few words at a time. Others get them right at the first attempt or never. The majority have occasional off-days but generally survive the ordeal without too much trouble. Relatively few have memories which allow them to recall prodigious numbers of words with barely a moment's hesitation.

For everyone else there is no infallible formula, certainly not ad-libbing, which is inclined to come across as uncertainty rather than spontaneity. Perhaps the only answer

Figure 9.2 One way out for the forgetful is to read a piece to camera direct from notes if you cannot remember the words or have no time to learn them. Have the memory-aid in shot, otherwise the audience may wonder why you keep looking away, and do not forget to look into camera as often as possible.

for the beginner is to keep the length of narrative down to the maximum capable of being remembered without difficulty. Anyone able to memorize a five-line limerick should certainly be able to recall 45 words (15 seconds). Trying to force more can be counter-productive, for little is worse than watching someone totter to within the final few sentences of a piece clearly too long to memorize.

While there is no way of avoiding the problem posed by the limits of memory, three possible escape routes suggest themselves. Using two takes framed in different ways

Figure 9.3 Another possibility for the forgetful. Anyone should be able to remember 45 words, so learn the two halves of a 30-second piece to camera separately. Get the camera operator to frame you in medium shot (a) for the first half and change to close-up (b) for the second. The pictures can then be edited together to produce one continuous take. Unsightly but effective. Note that the reporter in both shots is framed more centrally than in Figure 9.2 to avoid exaggerating the jump between edits.

(Figures 9.3(a) and (b)) is relatively ugly to look at and strictly second best, but, in a tight corner is preferable to a halting performance which threatens to grind to a full stop at any moment: Escapes 1 and 3 are suitable for use in recorded or 'live' conditions.

For Escape No. 1 the reporter needs to ensure the opening paragraph at least is word-perfect. The rest may be read from a notebook or clipboard which is clearly in shot so the viewer is not left wondering about the inevitable head movements away from the lens. Subsequent raising of the head from the page towards the camera for a sentence or two at a time may add just enough refinement to make the performance tolerable.

Escape No. 2 requires even more than usual cooperation from the camera-operator. Here the reporter does not attempt to speak the lines in a continuous take. Instead, the script is learned and recorded in two separate chunks of (again) say, 45 words. For the sections then to fit neatly together the two shots of the reporter must be framed in a sufficiently different way to avoid an awkward 'jump-cut' in the middle. The use of this technique should give the reporter some confidence early on. Later, attempts should be made to train the memory to accept longer and longer pieces, eventually overcoming the need for any split.

Escape No. 3 has been devised by reporters themselves. It involves pre-recording the words onto an audio cassette machine small enough to conceal in a pocket, and then listening to the replayed tape through a tiny earpiece while repeating the commentary to camera at the same time. It is an ingenious technique which seems to require more effort than that merely to learn the words, but practitioners say the recorders give them a confidence they would not otherwise feel, and if it improves their performance for the camera, then it is thoroughly worthwhile.

The method is not foolproof, though, for there is enough evidence on tape to prove that the all-important earpiece has a habit of popping out into sight at the wrong moment, and that cassette players have been known to malfunction and leave the performer speechless.

Finally, there is no point in any piece-to-camera script which fails to refer, even obliquely, to what is going on in the rest of the picture behind the reporter. When the background is general rather than specific, it is essential for script and location, however fragile the real connection between the two, to be tied together as firmly as possible, preferably by the opening words.

Studio spots

If the piece to camera is a vision story delivered on location, the studio spot is a vision story delivered in the studio by someone other than the main newscaster or presenter. Usually it is a specialist correspondent or reporter who is called upon to draw together the elements of a news story and tell it to the camera in the studio, often standing and with the aid of graphics on a massive 'video-wall' behind the journalist or to one side. Sometimes the correspondent also links in and out of clips of videotape.

For many reporters, these appearances aggravate the underlying nervousness already discussed, largely because of an awareness that such performances are invariably live ones made in the context of a programme as it is being aired, and that the smallest mistakes are therefore incapable of being corrected. That said, there are considerable advantages for those making studio appearances. First, the script can be prepared up to the time of transmission, making it possible to include the latest information about a running news story. Second, the performance is made in the reasonable comfort of

Figure 9.4 One of the most popular modern techniques for delivering pieces to camera. Read the written script into a miniature audio cassette machine and then replay the recording to prompt you as you speak the words aloud. It is, however, dangerous to try this for a live contribution, in case the audio cassette jams.

the studio rather than on location where other people and the elements make the situation harder to control. Third, and probably most important, the reporter may not need to rely on memory, for as well as the written script available out of shot, there will be a device displaying the words so that the speaker appears to be looking directly at the viewer while reading.

These devices, used by news presenters in particular, are often referred to by the general term 'prompter'. Various systems are used in television but they all simply take the words written into the computer and display them for the journalist as he or she looks directly at the camera lens. The main benefit here is that very late changes made on the master script are carried out simultaneously on the prompter. Where computer systems are in use, hard copy scripts are also likely to be printed out for readers who

Figure 9.5 The ultimate memory-jogger for reporters on location – a portable electronic 'prompting' device fitted to a hand-held camera. It is now just a matter of correct positioning to enable the words to be read. This is rarely used for daily news programmes but is more common for longer current affairs programmes or big live OB events. (Photo courtesy of EDS Portaprompt.)

do not wish to rely entirely on electronic viewing aids, and who like to mark emphasis, punctuation, and so on.

The rest of the prompter consists of a display unit mounted on the front of one or more studio cameras with the words superimposed on the lens photographing the performer, who looks straight ahead and reads at the same time. The scrolling movement of the script is regulated by an operator, who keeps pace as the words are spoken. Other computerized refinements allow the order of scripts to be changed at the press of a key.

Prompter devices are also used by politicians and other speakers who need to appear as fluent and at ease as are television performers in the studio.

All these machines, if used with care, are capable of producing fluently professional performances from almost anyone able to read with some expression. It is true that some reporters and programme presenters consider themselves above using any mechanical aids, but they are few enough to be regarded as unnecessarily eccentric. For most, the use of any sort of prompting device is infinitely preferable to the alternative: the sight of the top of the head during a performance because constant downward movement has to be made towards the script on the desk.

Prompters are as capable of being misused as any other tool, and many nervous television performers tend to depend on them as lifelines from which they dare not be parted. The frequent result is a near-hypnotic gaze which seems to bore into the viewer. The really skilful and experienced user of the machine treats it as a valued friend and ally, yet is sensible enough not to depend on it entirely. Occasional references to the written script, particularly for figures or statistics, reassure the viewer that details are not just being conjured out of thin air, although by now most members of the public must realize that no reader could possibly have learnt it all.

Prompting devices in general have proved a real boon to news performers, and have helped programme presentation to the extent where it is difficult to know how any modern television news service can afford to be without them.

Interviewing

Probably more indignation is aroused by interviews than almost any other aspect of factual television, although the objections may arise less frequently over news interviews than those which come under the loose heading of 'current affairs'.

Complaints tend to fall into three main categories. There is, first of all, the matter of intrusion where, instead of respecting the privacy of, say, the newly bereaved, interviewers are seen and heard callously asking questions apparently without a qualm. Any questioning in these circumstances would seem to be extremely difficult to justify in a civilized society, yet a surprising number of people are willing, anxious even, to talk about their tragic experiences, either as a form of mental release or as a genuine attempt to prevent similar misfortunes befalling others.

The ethical considerations implicit in such interviews have become the subject of serious study by academics and bodies such as Victim Support, who make eloquent pleas for treating with respect the people who (without wishing to) provide their subject matter.[1] Codes of conduct and other professional guidelines set out to govern the way journalists behave on a voluntary basis, but a number of notorious examples – including the publication of newspaper photographs showing well-known people in compromising or very personal situations – have led to the threat of some statutory restrictions on all the media. The defence of public interest is not always accepted by those outside the media industry.

Certainly some reporters are guilty of overstepping the mark by asking penetrating questions of those still too dazed to comprehend quite what they are saying. But such interviews have long been considered legitimate journalistic practice, and will no doubt remain so. The most that can be asked of television reporters is that they respect the privacy of those who wish it, put their questions with tact and sympathy, and do not demand answers as of right.

1. *Victims Twice Over* (1992). A report by Victim Support on treatment of victims of crime by the media.

The second category of complaint concerns the attitude taken towards the subject being interviewed. A substantial body of opinion exists which takes exception to any form of questioning probing beyond the simple elicitation of facts. This objection is usually framed as a demand to know 'How dare they ask ...' this or that, or what qualifies the questioner to appear to doubt what he or she is being told (especially when the interviewee is a respected or well-known figure).

At its worst, this technique has been termed 'trial by television', and has been most graphically illustrated by programmes, probably not quite so prevalent any more, in which interviewees have been browbeaten into making sometimes damaging personal admissions by relentless questioning bordering on interrogation. A secondary complaint about this kind of television is that the interviewers seem as much concerned with projecting their own egos as in producing serious answers to serious issues.

Perhaps it was only to be expected that the pendulum would swing so far the other way after the bland, deferential questioning general in the early 1950s, when interviewers seemed entirely content with any platitudes uttered by public figures, and shrank from querying the answers, however unenlightening they may have been.[2]

The precise moment at which a recognizable change of attitude took place is impossible to identify, although many television professionals are convinced it all happened

Figure 9.6 News in action. Reporters and crews form a 'scrum' around a subject, in this case the politician John Prescott. Scope for the carefully prepared question barely matters in these circumstances. Crews and reporters, who one moment before had been sharing friendly gossip, suddenly become competitors to get the best shots and the best sound. Questions are shouted out, usually short and simple to get a response. (Photo courtesy of and © Sky Television.)

2. Michael Cockerell (1989). *Live from Number 10: The Inside Story of Prime Ministers and Television*, Faber and Faber.

Figure 9.7 A watershed in British television news interviewing. Robin Day was one of the first journalists to treat interviewees more naturally, as equals and with no more deference than they deserve. In this incident in 1957 Robin Day challenged the Japanese Foreign Minister about Japanese engineering, producing some ball-bearings and asking why they resembled an identical British-made product. The interpreter, instead of just translating the question, complained that no advance notice of the question had been received and accused Robin Day of being disrespectful. (Photo courtesy of ITN.)

almost half a century ago. This was an interview between Robin Day and President Nasser of Egypt in Cairo for ITN in June 1957. Here was a head of state answering with apparent frankness and sincerity questions which diplomats would probably have hesitated to ask. Read today, the full transcript[3] makes the interview seem innocuous enough but at the time it created a real impact on the viewing audience and, unknown to the participants, set a standard for future generations of reporters.

In all the years since then, attempts have been made to strike a reasonable balance between a persistent, unshakeable seeking after truth and the asking of legitimate questions in a non-truculent manner. However, it is interesting to note that even in some Western democracies hard-edged political interviewing scarcely exists, and it is not unknown for government ministers to seek to influence not only the questions but who the interviewer will be. In Britain, the 'blackballing' of a television journalist's application for membership of a private club was thought to be not entirely unconnected with his famously combative interviewing technique.

3. Sir Robin Day (1993). ... *But with Respect*, Weidenfeld and Nicolson.

At the same time many present-day interviewees are as conversant as the interviewers with the rules of the game. Some are even better at it, because the questions they are being asked are connected with the subjects on which they are experts, and a reporter who is unsure of his or her ground for lack of briefing or any other reason can be made to look very foolish. Really experienced interviewees have the knack of being able to answer any awkward question in a way which suits them, and they can use their knowledge of studio routines and the programmes they are appearing on to make sure that there is no time for dangerous supplementaries. It can be an uncomfortable education to watch an experienced journalist being given the runaround by a wily interviewee who has been through it all before. One of the mistakes inexperienced interviewers can make is to ask two questions rolled into one – the clever interviewee will choose whichever point is easiest to cope with.

It was during the 1980s when interviewees decided they wanted to have their say and refuse to be intimidated by the cameras, studios and the bright lights of the television interview confessional. An early exponent of the art of hijacking an interview for his own aims was Sir James Goldsmith, the millionaire businessman. Goldsmith, who believed that journalists who specialized in business were excessively cynical about people who made money, agreed to a live interview on a BBC current affairs business programme. He totally ignored the questions and instead zipped into a detailed point-by-point criticism of a film they had made about his company. He eventually turned in his chair, produced a typed transcript of the film and threatened to read it out, live on air. Then there was Arthur Scargill, President of the National Union of Mineworkers. As leader of the long-drawn-out coal miners' strike of 1984–5 he became a pathfinder for other difficult-to-pin-down personalities, laying down the conditions under which he was prepared to do interviews and then, no matter what the question was, sticking doggedly to his prepared line. It was probably the first time since modern television interviewing techniques began that an interviewee exercised such domination over his questioners.

Some in public life go for coaching at television schools created by enterprising business-people, often ex-television professionals themselves. Would-be spokespersons of all types submit themselves to grillings intended to simulate as closely as possible the battleground of the television studio. What to wear, how to behave, whether to accept alcohol before an interview and similarly practical tips often come as part of the training. Other prospective interviewees equip themselves with audio tape recorders, which they place in prominent positions during interviews of all types for use as an insurance against being quoted out of context.

Some interviewers have raised objections to these methods on the grounds that they are designed only to provide the interviewee with enough skill to evade the difficult questions. But 'media training' is big business and here to stay and it seems unreasonable for the professional to quibble because the odds (which are nearly always stacked heavily against the inexperienced interviewee) should have become slightly more even. And since the whole object of the exercise is supposed to be to give the viewer a wider knowledge of what is going on in the world, anything which helps an interviewee to achieve that end ought to be welcomed.

The able, intelligent, well-briefed interviewer should always be capable of extracting something worthwhile in the course of reasoned argument. Where there is evidence of evasion, the reporter should not hesitate to pursue, although without bullying. That is ill-mannered, unnecessary, and always counter-productive, since viewers are much more likely to be sympathetic towards anyone they regard as the underdog. Besides, an evasion or point-blank refusal to answer speaks eloquently enough for itself.

The third main area of controversy surrounds the editing of interviews. Politicians and other public figures who appear frequently are well aware that in countries where editorial freedom exists, not all they have to say is going to find its way onto the screen, however important they consider it to be. Even where live contributions are concerned they know there is never likely to be enough time available to go through all the arguments, and are therefore content to try to restrict themselves to making those few main points they consider essential to their platform, in the belief that even minimal exposure to the public is far better than none. They have learned how to construct speeches in a way which directs the journalists to the juiciest sound bites they hope will be most likely to influence the audience. In pre-recorded interviews in the 1990s politicians and people regularly interviewed knew that only a 20 to 25 second clip of what they said would be used, and they responded with that clip in mind. But when 24-hour news services expanded they had to be reminded by their aides that an interview they thought was designed for a small, neat clip (or sound bite) could well end up being broadcast in its entirety on a news channel with plenty of airtime to fill.

It is usually from the interviewee inexperienced in the ways of television that the loudest cries of 'misrepresentation' are heard, and there can be only sympathy for those whose parenthetical comments are construed as the real substance and are extracted for use from a recorded interview. In all cases, reporters should make it clear that they are under no obligation to use the whole of an interview, or indeed any of it, and give the subject an opportunity to make out a case in a cogent fashion or withdraw altogether.

Neither should any doubts be left about intended usage. An interviewee has the right to be told of the context in which the contribution is being made, and whether contrary opinions are being sought on the same subject. For although it would be intolerable for journalists to lose the freedom to edit as they think fit, they should be equally jealous in guarding the rights of their interviewees to fair and honest treatment.

Purpose and preparation

Although it is right for those who ask the questions sometimes to be taken to task by their subjects and other critics, it could also be argued that the editorial imperative for conducting interviews is not always thought through in anything like enough detail.

For many programme executives in television news – and radio, with much more airtime to fill, is probably an even greater culprit – the knee-jerk response to a news issue is 'let's talk' to someone. It is often easy to understand why. For instance, a 'live' interview has the benefit of being shortened or lengthened during the course of a newscast (a godsend to a hard-pressed editor sweating over the likely appearance or non-appearance of items still being edited for transmission). Yet when the interview has been transmitted, perhaps after considerable effort and expense has been spent on ensuring that the 'someone' is brought within range of the reporter/presenter and camera, a certain vague dissatisfaction with the product may linger. Chances are that in that case three fundamental points were overlooked by the editorial team when the interview was being commissioned:

1. Is an interview the best way of proceeding?
2. If so, are we talking to the right person?
3. In that case what do we hope will come out of it?

Quite often, 'talking to someone' is simply not appropriate for the story, which can be told more succinctly by the reporter on the spot or the presenter in the studio.

Then comes the question of whether the right person is being approached. Critics of television news grumble about favouritism on the part of editors, and hint at a conspiracy to exclude all but a small band of 'resident experts' they have found to speak on various subjects, and whose views are sought regularly and at length. There is no conspiracy. Broadcasting demands its contributors to be articulate, fluent and concise, so it is not surprising that editors prefer the security of performers whose reliability is proven rather than gamble with the untried and step into the unknown. Journalists themselves may rail against 'rent-a-quote' politicians and officials, but the onus is on journalists themselves to widen the net. It is fair to say that considerable efforts have been made in some quarters to draw in a more representative sample of 'expert' interviewees, women and members of the ethnic communities in particular, but although this approach is to be applauded, going to interview the wrong person for the right reason is counter-productive.

For many interviews, however, the weakest element of the formula is the last. Having decided that an interview is required and the most appropriate person found, reporters are too often unsure what they want from them. 'What shall I ask you?' is a question I have heard put anxiously by interviewers, meeting their subject for the first time a few minutes before transmission. It is not a practice likely to inspire confidence. Most interviewees are fully aware that they know more about the topic than the person asking them about it – after all, that is why they are there – but intelligent questions are based on background knowledge, however limited.

The idea is therefore not to begin an interview unless or until some inkling of what will come out of it is clear. Opinion, interpretation or explanation, certainly, but based on proper preparation and not through 'research' gathered on screen during the course of the interview.

It may seem a ridiculously obvious thing to say, but reporters must know whom they are questioning. Professional journalists are not immune to nervousness or momentary lapses of memory, so confirm the details in advance and write them clearly on a pad or notebook for immediate reference. In this way it will be possible to avoid the acute embarrassment of having an interviewee preface his or her first answer with a correction of name, rank or title.

Four main forms of news interview

Watch any conventional television news programme and it will quickly become apparent that a great deal of the speech coming from the mouths of contributors is in the form of answers to questions: in other words, interviews. Of course, these may be of such minuscule duration that they barely deserve to qualify for the term, but the point is that for a soundbite answer to be selected, someone has to pose a question in the first place.

Interviews also probably represent the highest proportion of contributions reporters make to news programmes, even though the credits may be given collectively ('We asked the Foreign Secretary ...') rather than individually ('Jane Smith asked the Foreign Secretary ...').

Of all the different types seen on the screen, possibly the one in most frequent use currently is the set piece, conducted on the interviewee's own territory, in some other appropriate setting, or 'down the line' from the news studio, for live or recorded trans-

Figure 9.8 Forms of interview. Top: Doorstepping, informal attempts by journalists to extract something worthwhile from newsworthy events. Middle: eyewitness interviews with those able to provide immediate information about events. Bottom: news conferences are a way of dealing with large numbers of questions.

mission. The important thing about the set piece is the presumption it makes of the interviewee's willingness to participate, allowing for arrangements to be made far enough in advance for the reporter to do some proper homework, including the preparation of questions.

Such luxuries are not normally afforded reporters assigned to eyewitness or spot interviews, where the most important journalistic quality in demand is the speedy rooting out of those who are willing and able to talk about the experience of events which may just have occurred. The questions here are more likely to concern facts rather than opinions.

Much the same can be said of the 'doorstepper' interview. This is the form of questioning most loathed by critics of the press and broadcasters who regard it as akin to a journalist burgling someone's home. They might be interested to know that the reporters hate it too. The reporter waits, perhaps literally on the doorstep of a building (hence the name), in order to snatch a few words, any words, with participants in some newsworthy event. These hit-or-miss affairs have become familiar sights on many a nation's television screen. In the general scrum all too often associated with these matters, the television reporter may get no further than thrusting the microphone forward to ask such elementary questions as 'What's been happening?' or 'What's next?' in the hope that what may start out as non-committal, even grudging answers may, with perseverance, become proper interviews in which some real information is forthcoming.

Even the refusal to say anything other than 'no comment' is often considered worth screening if only to show the manner of rebuff. Some seasoned interviewees are not above turning the doorstep or shouted question to their own advantage. There were recent American presidents who managed to reply to some questions shouted above the noise of that helicopter landing on the White House lawn, but strangely, found themselves unable to hear the more awkward ones. Bill Clinton, and George Bush before him, frequently developed this hearing problem.

This method of interviewing is one stage removed from the vox pop (*vox populi*), an entertaining if often inconsequential sounding out of opinion among people, members of the public, stopped in the street. The aim is to achieve a cross-section of views or reactions to a specific topic, with each contribution usually boiled down to a few words. The technique calls for the same questions to be put in the same way each time so that the answers may be edited together without the interviewer popping up in between to spoil the flow. They are often in the 'Do you prefer cats to dogs?' category, although local news programmes often reflect reasonable public views about more serious subjects, such as a local road scheme or the closure of a factory.

All interviews may be said to be variations on these four broad categories, including the news conferences so beloved of politicians and others. These occasions are usually held to avoid the need for separate interviews, and so take place somewhere large enough to accommodate all the television, radio and print journalists who wish to attend. The numbers are likely to run into hundreds, presenting enormous difficulties for the television news reporter, who may not get the opportunity to put a single question.

At other times, paths may be smoothed by the organizers' imposition of an advance order of questioning, although this can be taken to extremes, with clever stage-management to ensure that there is time for only 'friendly' questions to be taken.

Settings for news conferences, as for many other newsworthy happenings, take the choice of location entirely out of the hands of the reporter and camera-operator. Then it is a matter of making do with whatever site is available, relying on good picture composition and sound quality to make the result as satisfactory as possible.

Far more desirable from the reporter's point of view is the selection of a background appropriate to the story being told. For example, it makes sense to interview the scientist in the laboratory rather than in front of a plain office wall, to talk to the newspaper editor against a background of newsroom activity, to the shop floor worker on the shop floor, and so on. Relevance should always be the aim.

One word of caution. There is a very fine line between a background of interest and one so absorbing it distracts the viewer from what is being said. Being too clever can also create unexpected problems at the cutting stage. I once chose to locate an interview with the managing director of a motor company on the brow of a hill overlooking the firm's test track. It was certainly relevant, the cars making a very pleasant sight as they whizzed across from one side of the picture to the other behind the interviewee's back. Unfortunately, it dawned on us only when reviewing the pictures afterwards that one particularly important editing point coincided with the moment when a car was in motion halfway across the screen, and that a straight edit into a later section of the interview had the effect of making it vanish into thin air. Also, never do an interview with a clock in the background – for the very same reason of continuity.

Although composition of the picture is the responsibility of the camera-operator, it is a poor reporter who allows the subject to be so badly framed that trees or other obstacles appear to be growing out of the interviewee's head. On one occasion a serious interview was reduced to near farce when it was discovered, too late, that what was an apparently ideal background of ornamental swords hung horizontally on the wall behind the interviewee had wholly unexpected results when seen on the two-dimensional television screen. A long, heavy blade appeared to protrude from the side of the subject's neck, and what made it all the more fascinating was that he did not seem to notice.

Putting the questions

Most journalists have had considerable experience as interviewers before they move to television, but there is a vast difference between the casual questioning which takes place in the quiet corner of a pub or over the telephone, and the paraphernalia of camera, lights and technicians.

The newspaper journalist is able to phrase the questions in a conversational, informal manner, interjecting now and again to clarify a point, jotting down answers with pen and notebook, or simply sitting back to let a portable audio recorder take the strain. Questions and answers need not be grammatical or follow a logical pattern. The same ground may be gone over and over again. If either participant has a cleft palate, stutter or some other speech impediment, no matter. The printed page on which the interview appears does not communicate that fact to the reader without the writer deliberately choosing to do so.

The passage of time and the advances in technology have done nothing to change the belief that in television journalistic judgement and writing ability alone are not enough. Another piece of advice from the past that has not changed came from Sir Geoffrey Cox, editor of ITN back in the 1950s. He identified the significant differences between the two forms of interview. It remains as true now as it was then:

> The best newspaperman will often take plenty of time stalking around his subject, taking up minor points before he comes to his main question, noting a fact here, or an emphasis there, and then sifting out his material later when he sits down at the typewriter. But the television journalist is forced to get to the point at once, as bluntly and curtly as is practicable. His questions must also be designed to produce compact answers, for although film can be cut, it cannot be compressed.[4]

Another, still relevant, piece of advice came from Michael Parkinson, an experienced practitioner in both newspaper and television interviewing. He put it rather more force-

4. *The Daily Telegraph*, 7 July 1958.

fully, describing the newspaper or magazine interview as 'child's play', compared with that for television. The modern generation of journalist-interviewers, called upon increasingly to display their talents live into newscasts, would no doubt agree with his broad sentiments:

> A three-hour chat over lunch, a carefully written, honed and edited piece and the journalist has created something beyond the reach of any interviewer on television who tries to do his job without the luxury of being able to shape his material after the event. One is instant journalism, the other retrospective. It is the difference between riding bareback and sitting astride a rocking horse.'[5]

Not every television interviewer would put it as baldly as that, but it is undoubtedly true that the screen interview of any type makes considerably more demands on the person conducting it. The essential qualities include an ability to think quickly to follow up topics outside any originally planned structure of the interview, and a capacity to marshal the thoughts in a way which builds up logical, step-by-step answers.

Each interview, however brief, is capable of taking on a recognizable shape. Questions which are sprayed in all directions as topics are chosen at random make the live interview difficult to follow and the recorded one doubly difficult to edit intelligently. In any case 'the office' would much prefer to select a chunk of two or three questions and answers which follow a rational pattern. Apart from that there is the waste of time involved in asking questions which have no real relevance to the occasion. The same goes for any attempt to produce a relaxed atmosphere by lobbing one or two innocuous 'warm up' questions at the beginning. Tape may be cheap and reusable, but no useful journalistic purpose is served.

The phrasing of questions also needs to be considered carefully. Too many inexperienced reporters, rather fond of the sound of their own voices early on in their careers, have a tendency to make long, rambling statements barely recognizable as questions at all. Aping the established 'names' whose deliberately long-winded approach is their trade-mark is not to be encouraged. Nor is the 'portmanteau', in which two or three questions are wrapped up in one. Inexperienced interviewees will be unsure which to answer, while seasoned campaigners will seize on the ones which suit them best and ignore the rest.

Much the same principle applies to broad questions of the 'meaning of life' variety. General questions tend to produce lengthy, wide-ranging answers, so it is better to focus on and follow through a narrowish line of questioning.

At the other extreme are the brusque, two- or three-word interjections which, apart from anything else, do not register on the screen long enough if faithfully repeated as cut-aways.

Next come the cliché questions. My favourite remains

> 'How/what do you feel about ...?'

a question which almost cries out for a rude answer. Others which crop up all too frequently include:

> 'Just what/how serious ...?'
> 'What of the future ...?'

5. *The Sunday Times*, 15 December 1974.

Then there is the tendency to preface virtually every question with some deferential phrase or other which is presumably meant to soften up the interviewee:

'May I ask ...?'
'Do you mind my asking ...?'
'What do you say to ...?'
'Could you tell me ...?'
'Might I put it like this ...?'

each of which invites curt rejection. Shooting straight from the hip has its drawbacks, of course. Without proper care, questions which are too direct are quite likely to produce a simple 'yes' or 'no' without further elaboration:

'Is it true you've resigned because of a personal disagreement with the Prime Minister?'
'Is there any chance you might return to the government in the future?'
'Have you decided what you are going to do now?'

If the television news interview is to be of any value at all, the questions must be constructed more skilfully, in ways designed to draw out positive replies:

'What do you say to reports you've resigned because of a personal disagreement with the Prime Minister?'
'How would things have to change before you would consider returning to the government?'
'What are you planning to do now?'

Be sure, though, you are on safe ground when using such phrases as '... reports you've resigned because ...'. Too many interviewers treat *people-are-saying* media speculation as fact and incorporate it into questions which are easily brushed aside. What they really mean is 'I think ...' but lack the knowledge, or the 'bottle', to say so.

As for general demeanour when asking the questions, interviewers should not allow themselves to be overawed in the presence of the important or powerful, or overbearing when the subject of the interview is unused to television. As (Sir) Robin Day put it in a ten-point code of conduct for interviewers he first suggested in 1961 and found just as relevant nearly 40 years later:

> a television interviewer is not employed as a debater, prosecutor, inquisitor, psychiatrist or third-degree expert, but as a journalist seeking information on behalf of the viewer.[6]

Sometimes, in seeking that information from an agreeable and fluent interviewee, it is tempting to try out the questions in a 'dry run' without the camera. That is a mistake. No journalist should compromise him or herself by submitting questions in advance, unless that is a condition of agreement to an appearance. Second, an interview based on known questions is almost certain to lack any feeling of spontaneity. Third, even the most apparently loquacious people are inclined to 'talk themselves out' during a formal

6. (Sir) Robin Day (1962). *Television, A Personal Report*, Hutchinson; (1993). *... But with Respect*, Weidenfeld and Nicolson.

try-out, becoming tongue-tied when the real thing begins. So while a brief discussion about the general scope of an interview is a sensible preliminary, any full-scale rehearsal should be eschewed.

Coping with the answers

It is all very well for the reporter to ask the questions the average member of the public would dearly love the opportunity to put. Actually coping with the answers poses a problem by itself.

The experienced politician is quite capable of turning aside the most difficult question with a disarming smile and a reference to the interviewer by first name. The others who are adept at ignoring questions and going on with their prepared answers can be 'chased' with repetition until they are caught or their evasions become obvious. But the reporter's real troubles begin when he or she does not listen to the answers.

This is by no means as uncommon as it may seem. The pressure on a questioner conducting an interview can be almost as great if not greater than on the interviewee, and it is all too easy to concentrate on mentally ticking off a list of prepared questions instead of listening, poised to follow up with an occasional supplementary. If the reporter's concentration wavers, any number of obvious loose ends may remain untied. Ideally, it should be possible to forget the camera and the rest of it, relying on sound journalistic instinct to take over.

Jumping in too soon with a new question before the interviewee has finished could seem rude to the subject as well as producing an ugly overlap of voices which may also be uneditable if the interview is being recorded. When an interruption is necessary, it is better to wait until the subject pauses for breath or the inflection of the voice is downwards. If the question is genuinely misunderstood, the reply halting, gobbledygook or off course in some other way, the reporter is well advised to call out 'cut' to stop the camera, so the problem can be discussed before restarting the interview. This is far more sensible than stumbling through to the bitter end and hoping there will be enough time left to try all over again.

The set-piece interview, step by step

As an example of the way a typical set-piece interview might be conducted, consider the sequence of actions taken in this imaginary question and answer session in the third-floor office of a government economics adviser. Assume that both participants have been adequately briefed. Time is important, as the reporter has an early-evening programme deadline to meet, and the adviser has to leave for the airport within half an hour.

Step 1: setting up

The office is of medium size, the furniture consisting of a large desk with a swivel chair behind it, one easy chair and a coat stand. The only natural light comes from a single window overlooking a main road and heavy traffic. On this occasion the camera equipment, having been manhandled from the lift a few yards away, now litters the floor. The camera-operator unstraps the legs of the tripod and sets up opposite but slightly to the left and in front of the interviewee's chair about three metres away. The camera

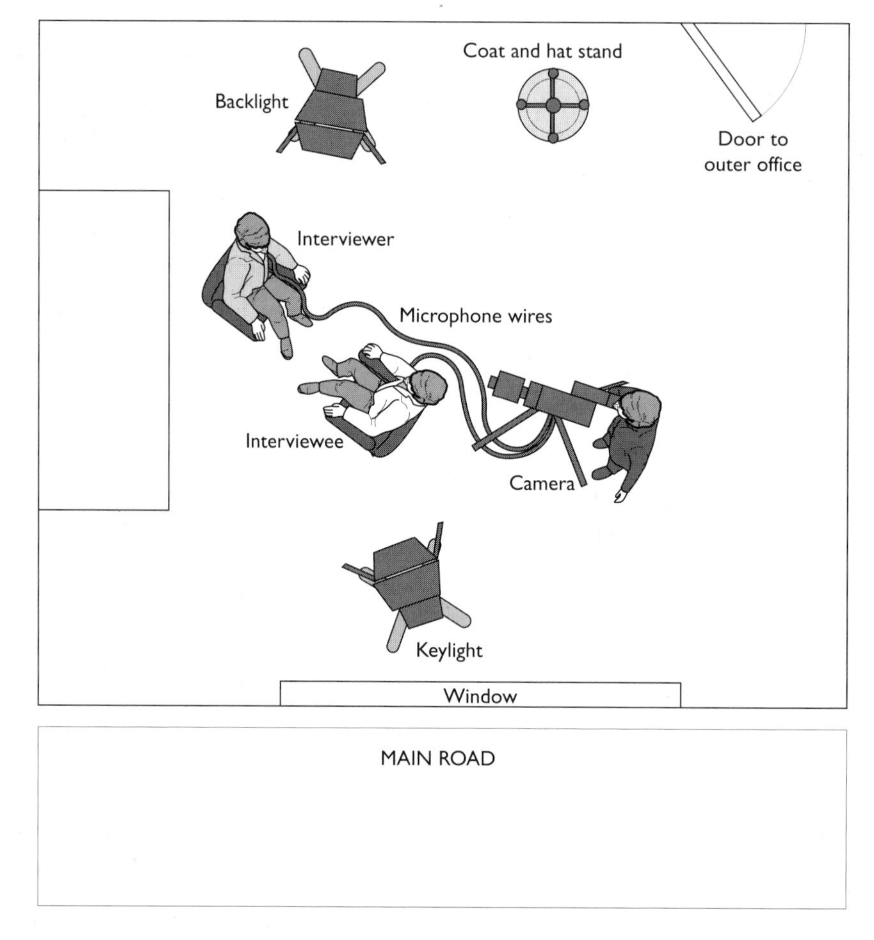

Figure 9.9 Typical office setting for a one-plus-one set-piece interview for television news.

is slid on to the base plate fitted to the tripod head and locked firmly into place, the camera-operator consulting the built-in spirit level to ensure that the camera is level. The reporter gets a chair from an outer office and places it about a metre to the left and in front of the camera. Having found the office power socket outlet, the camera-operator puts up two 800 watt lights. Both are on stands which are extended to their full height so the lamp heads can be angled towards the participants without blinding them. One becomes the 'key' light behind the reporter's chair, the second, the back light, is positioned out of shot, over the interviewee's shoulder, 'barn-doored' to keep the light off the camera lens.

This, then, is the classic interview set-up, with the camera in position as the observer of a discussion between two people. But as it will concentrate entirely on the interviewee, compensation has to be made for any impression that no one else is in the room. For this reason, as the lens looks over the reporter's shoulder, it frames the subject in three-quarter full face, slightly off-centre, making him look across the empty side of the screen towards the questioner. Moving the camera more to the left would have the interviewee looking directly into the lens, suggesting a party political broadcast or some sort

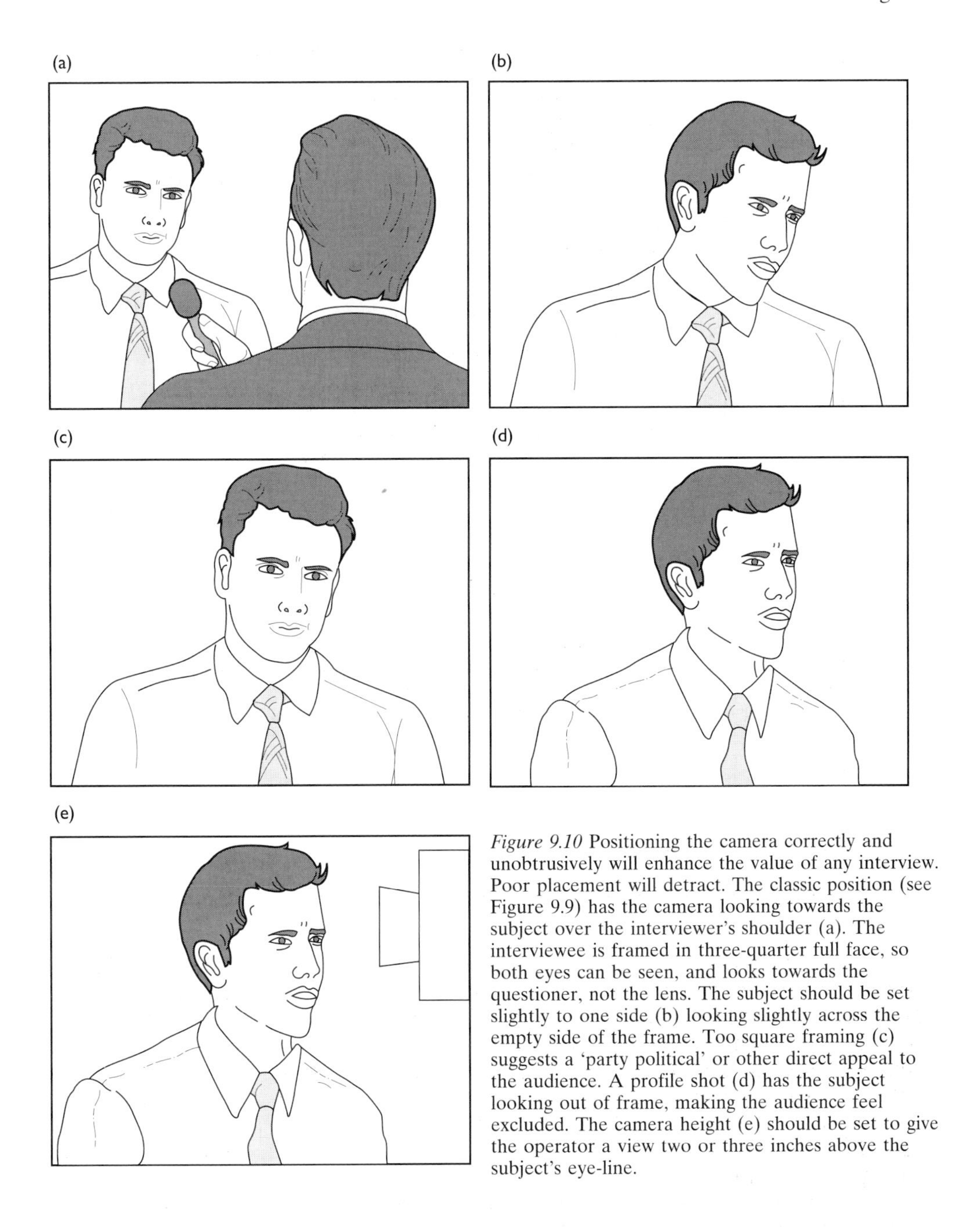

Figure 9.10 Positioning the camera correctly and unobtrusively will enhance the value of any interview. Poor placement will detract. The classic position (see Figure 9.9) has the camera looking towards the subject over the interviewer's shoulder (a). The interviewee is framed in three-quarter full face, so both eyes can be seen, and looks towards the questioner, not the lens. The subject should be set slightly to one side (b) looking slightly across the empty side of the frame. Too square framing (c) suggests a 'party political' or other direct appeal to the audience. A profile shot (d) has the subject looking out of frame, making the audience feel excluded. The camera height (e) should be set to give the operator a view two or three inches above the subject's eye-line.

of appeal. More to the right would produce a full profile, making him look out of the picture – television's equivalent of allowing the subject of an end-column newspaper photograph to look out of the page. The height of the camera is set so that the camera-operator's view is 50–80 mm above the subject's line of sight (eyeline). Going below the eyeline and looking up has a distorting effect.

Step 2: final preparations

The two main characters emerge from an outer office where they have had a brief talk about the area the interview is intended to cover, although the reporter has been careful not to go into detail about the questions. As they settle into their seats, the camera-operator attaches a tiny microphone to the clothing of each person. Now, with the recording cable plugged into its socket at the back of the camera, the camera-operator asks each participant in the interview for a burst of 'level' to ensure that both voices are heard at equal strength. Traffic noise from the street three floors below is minimal thanks to the window blind, but the sound of the telephone ringing intrudes through the thin office wall. It is probably the reporter who asks the secretary next door to ensure that any calls which would interrupt are diverted to another extension.

The reporter refers again to the questions scribbled on his notepad and switches on the tiny audio cassette recorder he has brought along. The camera-operator checks that a fresh video cassette has been inserted into the camera, switches on to the VTR stand-by position, gives the focus on the 14:1 zoom lens a final tweak, checks the automatic exposure, and takes one last look around the room to ensure that everything is properly in position.

Step 3: interview in progress

'Quiet please', calls the camera-operator. A second or so later he flicks the switch to start the recording, checking to ensure that the tape is beginning to roll from the cassette. For the first six or seven seconds he records what is known as an establishing two-shot, taking in the back of the reporter's head, the interviewee and the top of the desk, to show the respective positions of the two people taking part. As long as the interviewee has been warned not to speak at this stage, the two-shot can be dropped in by the picture editor over any of the questions, or used as a means of joining separate chunks of interview neatly. This shot complete, the camera-operator now frames the interviewee on his own, calling out 'action', 'running', or simply 'go ahead' as a cue for the reporter to begin asking his questions.

As a typical news interview is rarely longer than six or seven questions, the camera-operator goes through his usual sequence of shots, using the zoom lens to give a pleasing visual variation and to aid swift editing later on. He does this without stopping the camera or the interview, but makes sure any movement is made only during the questions, otherwise editing would be more difficult. He holds the first two questions and answers in a steady medium shot, from the waist up. As the third question is being asked he tightens to a medium close-up of the head and shoulders.

At this point, with the digital time-code display on the camera showing that 6 minutes 45 seconds of tape have been used, the reporter calls out, 'Cut!' The camera-operator switches off, and turns off the lights which have made the room rather warm. The interviewee is disconnected from the microphone lead, shakes hands all round, excuses himself and prepares to depart for the airport roughly ten minutes earlier than he expected.

Step 4: cut-away questions

Now comes the time to complete the cut-aways or reverses. Because only one camera was being used, the shots during the interview were concentrated on the subject, so the reporter now has to repeat his questions for possible use by the picture editor. It is a method of bridging different sections and inserting the interviewer in a way which looks

Figure 9.11 Varying the shot. A straightforward news interview is likely to be based on three pictorial elements: a two-shot of both subjects to set the scene and a medium close-up and close-up of the interviewee. 'Reverse angles' of the interviewer will be shot later (see Figure 9.12).

smooth on the screen, but it is a device which has fallen into disfavour with some news services, which prefer to 'jump cut' between portions to show that interviews are edited.

Cutaways are carried out in two ways: the reporter can play back the conversation as recorded on his or her own audio machine, or ask the camera-operator to rewind the tape and replay interview sound and picture through the camera viewfinder. The aim in either case is to ensure that the questions follow both the phrasing and tone of the original. On this occasion the reporter settles for his audio tape recorder, and jots down

Figure 9.12 Where interviews are conducted with a single camera, as on most news stories, reverse angle shots are sometimes used to 'insert' the reporter at the editing stage as a way of bridging sections neatly. Correct repositioning of both camera and interviewer is essential, or the questioner may appear to be looking in the same direction as the interviewee and not towards them. The camera operator's rule: the ear the camera sees during the interview (a) must be the same one the camera sees in the cut-away (b).

the information he needs alongside some of the original notes he made as a guideline before the interview.

The camera-operator now has to concentrate on the face of the reporter, which he was unable to do during the course of the interview, and it means moving either the subject or the camera to a new position. The camera-operator concludes that in this case it is simpler to ask the reporter to move his chair and to adjust the lighting accordingly. Now he re-focuses, taking great care to ensure the eyeline is correct by keeping the camera at the same height as before. To help the picture editor put the interviewer accurately into the edited story, the camera-operator follows the simple rule he made for himself years earlier: the ear the camera saw during the interview proper must be the same ear the camera sees in the cut-away questions. Getting it wrong would create an impression of interviewer and subject looking in the same direction instead of at each other.

As an extra insurance, the interviewer is framed throughout in a medium shot, so wherever the cut-away appears the reporter will seem to complement and not dominate the subject. The questions themselves follow the original sequence. Although they are now being asked of no one the reporter tries hard to recreate the spontaneity of the interview which took place a few minutes earlier, taking the opportunity to tidy up one or two insignificant grammatical errors. After each of the cut-aways he looks down at his pad to refresh his memory, pausing deliberately for two or three seconds before asking the next question. He knows if he speaks while still looking down at his notes the picture editor will be unable to use the shot.

To finish, the reporter does several 'noddies', understated movements of the head to simulate reactions as they might have occurred at various points during the interview. These serve the same purpose as the questions, acting as bridges between edited sections, but reporters generally seem unable to carry them out with much conviction, and so the technique fell into disuse in the late 1990s. There is also a feeling among newspeople that in some interviews the inclusion of a 'noddy' might create the impression that the reporter was agreeing with what was being said. As an alternative, camera-operators sometimes record 'steady listening shots', of an attentive but otherwise immobile reporter and, when time allows, a close-up listening shot of the subject is also added to give the picture editor maximum choice of material.

Studio interviews

Although it is possible to record or transmit live a typical studio interview between two people using two cameras, or at a pinch only one, most studio directors would prefer to use three, one concentrating on each participant and the third to provide the variety of a two-shot.

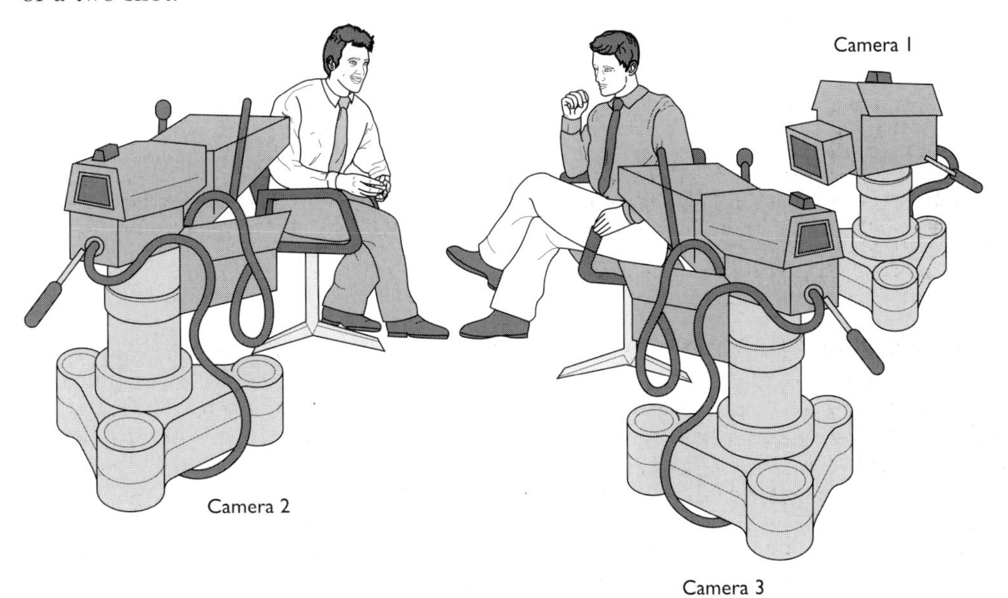

Figure 9.13 Live studio interviews in news programmes offer flexibility and the opportunity for more sophisticated production than can be achieved using a single camera on location. Many studio directors will have access to three cameras – one concentrating on a medium close-up/close-up of the interviewer, one concentrating on a medium close-up/close-up of the interviewee, and the third for a two-shot. With careful planning it would also be possible to move cameras one and two during the interview to provide pleasing 'over the shoulder' variations.

As parts of newscasts rather than as programmes in their own right, live interviews have the merit of being malleable. Those which are not getting anywhere can be savagely shortened. Good ones can be allowed to continue at the expense of other items dropped during transmission. Then there are some producers who, without much regard to quality, will happily tailor their interviews to fit whatever odd spaces remain in their programmes, and there are still more who have cause to be thankful for the interviews which are able to fill a void created by a sudden breakdown in another technical area.

The satisfaction of those conducting any one of these battles of wit is that the outcome depends on the judgement and ability of the interviewer alone, and has to stand without the benefit of editing after the event. The techniques required are generally no different from those used in other types of interview, except that the questioner probably needs to be more mentally alert than ever, and deeper briefing and preparation are necessary to ensure there is no drying up.

The one extra skill which has to be developed is a sense of timing. The interview must not be so hurried that there is nothing left to say and still some of the allocated time to spare. Neither must there be such dalliance over the first part that important ground remains to be covered and the interviewee has to be cut short in full flight.

Various timings are given by the studio gallery as the interview proceeds, the most crucial time being the final minute. The interviewer has an ear tube through which timings are provided by the director, the programme editor or the production assistant. The final ten seconds are counted down separately. On zero, it is time to say something like: 'Well, we must end it there.' The most experienced interviewers are really adept at this timing exercise, able to wring the very last ounce of value from their interviews right up to the dying seconds.

Modern studio techniques may also link the studio director (and through him or her the programme editor/producer) in the control room to the interviewer in the studio through the earpiece. In this way instructions about timing as well as editorial matters can be given direct. Some presenters and reporters like to use this system throughout live newscasts, preferring the babble in the control room to the isolation they would otherwise feel. They say they are not distracted by the general background noise and can easily pick out instructions meant for them.

Figure 9.14 'Down-the-line' interviewing, linking the journalist in the studio with an interviewee elsewhere, has become a routine element in many television news services.

The growing popularity of 'live' news interviews has seen a corresponding increase in a production technique which keeps the interviewer in the studio while the subject is elsewhere. The main difficulty this 'down-the-line' interview style presents is that more often than not the journalist is unable to see the person being interviewed, even though their image may appear to the viewer to be in shot. This is done electronically, and means the interviewer has to turn to face wherever the subject seems to be.

Making any interview seem realistic while looking at a blank studio wall can be very awkward for the inexperienced to master, especially without any rehearsal. There are occasions when extra monitors can be placed on the desk to enable real visual contact to be established (although the interviewee is very unlikely to be able to see the questioner) but this calls for the setting up of additional shots on the studio cameras and at some stage during the newscast the equipment will probably have to be removed.

10

Packaging the news

The test of the television reporter's talents comes when separate news-gathering and production skills have to be fused together in the form of a news 'package', an all-picture report which appears on the screen complete in itself apart, invariably, from an introduction read by the presenter in the studio. The package will combine clips of interviews, good natural sound, the most appropriate pictures and the reporter's own voice (usually called the voicetrack). Often graphics will also be included if the report has complex elements which require further explanation, or numbers and statistics.

Elsewhere in television journalism, the likelihood is that editorial responsibility for overseeing something so complex will be given to a producer or director, but in modern news, with its slim-line approach, adding another body to the team is simply not an option. News programmes of varying lengths have come to rely on 'packaging' as their staple diet, and expect reporters to provide it, often judging them on their ability to produce consistently effective examples of the genre. Live news also requires a package as a kind of introduction to a subject and is usually followed by several live interviews, either in the main studio with the presenter, in a separate studio, or both. This is most common in programmes like *Channel 4 News* or the BBC's *Newsnight.*

The successful package calls for the aptitude to tell a story – not only the performing talents so apparent in the piece to camera or the stubborn persistence needed in some interviews. It demands conspicuous planning and organization, the ability to fit different pieces together to produce a coherent, continuous narrative, and a mastery of words and pictures. When all that is accomplished, the report must be capable of being assembled quickly, or enough information communicated to ensure it can be put together by other people, no matter how far away they are.

Packages do not have to be complicated just because they are invariably of substantial duration. Experienced reporters and camera teams, working closely towards a common goal, are well aware of the dangers of wild over-shooting and unnecessary complexity, while at the same time they ensure coverage is not skimped.

Broken into its separate elements, a typical package might appear to the viewer in the following order:

1. an opening picture sequence accompanied by the reporter's out-of-vision commentary;
2. an interview;
3. more pictures and commentary linking into ...
4. a piece to camera.

The chances are that the sequence in which they were completed was very different.

Some news services like to stamp their own individual styles on package formats, although these preferences might be only as straightforward as insisting they always begin with the reporter's piece to camera (to establish presence and authority at once), or never ending with an interview, expecting the reporter to complete the report with a visual pay-off.

But these are really minor considerations. From the reporter's point of view, the versatility of the package technique ranks high among its most satisfying aspects, with the separate ingredients capable of being cemented together in any number of ways, according to story demands. The one proviso is that during shooting reporter and camera-operator keep the eventual shape of their item firmly in mind, otherwise there is a strong possibility that there will be too much or too little material to go with a particular sequence. The latter fault may make it impossible for the picture editor to assemble the report without blatant disregard for visual grammar, or else become involved in the difficult, time-consuming, reporter reputation-wrecking operation known disparagingly as a 'salvage job'.

'Topfield' picture package

For an example of the way a typical picture report might be constructed, take this routine, imaginary piece about the increasing economic pressures on small dairy farmers trying to make a living against European competition. Assume the item has been commissioned by the editor of a programme we will call *The Five O'Clock Report* to accompany a story about a meeting of EU agriculture ministers the same day. Arrangements will have been made well in advance to secure the cooperation of the farm owner who, while not anxious to publicize a predicament which could lead to the loss of his prize herd of Jersey cows, is prepared to represent others in a similar plight.

Although an unrepentant 'townie', Delia Ward the reporter, who is something of a specialist in business and industry matters, has briefed herself well enough on the subject of farming to be aware of the general ground the assignment is meant to cover. A short talk with the farmer immediately on arrival fills in the necessary detail. The camera-operator, meanwhile, takes the opportunity to spy out the land without the encumbrance of his equipment, and is already impressed with the potential for vivid pictures.

Friendly, helpful and articulate, the farmer nevertheless has a number of important duties to perform about the place, and is anxious to complete his own part in the proceedings as quickly as possible. The first ingredient, therefore, must be the interview, which the reporter and camera-operator originally envisaged taking place later on, set against a background of the prize herd being milked. But the farmer's declared intentions have now forced a change of plan so, too early for the milking, they settle for an exterior view of the shed, with half a dozen milk churns clearly visible. Six questions and answers follow smoothly, rounded off with the obligatory two-shot and cutaway questions, all completed with under four minutes of tape on the digital counter. The reporter and camera-operator, now left to themselves following the farmer's departure, decide after a brief discussion to concentrate on other buildings, activity, people and animals around the farm, working from the perimeter in towards the farmhouse. It is an attractive building and the camera-operator takes several shots of the exterior from several angles.

Next they move inside to the kitchen, where the sunlight filtering through the big windows is so strong the camera-operator needs no artificial light for shots of the

Figure 10.1 Anatomy of a package.

Shooting sequence
1. Interview with farmer
2. General shots activity (1)
3. Piece to camera
4. General shots activity (2)
5. Recording of 'wild' sound
6. Recording of voice-over commentary
7. Final shots activity (3)

Edited sequence
1. Voice-over commentary
2. Interview
3. Voice-over continued
4. Piece to camera
5. Voice-over commentary pay off

farmer's wife making her account books up to date. All that takes another two-and-a-quarter-minutes of tape.

Over a break for coffee, the reporter – now able to imagine how the rest of the item will go – writes and memorizes the script for a piece to camera, a little over 30 seconds summing up the dilemma for farmers of this type. Outside again, she asks to be framed in a fairly loose medium shot against the background of a haystack to emphasize a point about increased feeding costs. She is virtually at the end of the piece when an unseen tractor starts up noisily, drowning her voice. A second take is necessary. This is completed satisfactorily and so far less than eight minutes of tape have been used.

The camera-operator is just removing the camera from the tripod, ready for the move to a new location, when he becomes aware of activity in the nearby lower grazing field. Milking, which he did not expect to take place for at least another fifteen minutes, is clearly imminent. In fact, the herd can be seen already beginning to move. Slightly annoyed at the unexpected need to rush, the camera-operator grabs his camcorder and makes off at a brisk trot in the direction of the advancing herd, the reporter in pursuit. There is no time to put up the tripod, so the camera-operator balances the camera comfortably on his shoulder and points the built-in microphone towards cows and farmhands on the move. Within a few minutes he has recorded onto the tape a series of close and medium shots, together with the ambient sound for use as effects. Then, holding the camera by the handle, he sprints to a conveniently elevated vantage point to capture the long shot of the procession filing past on the way to the milking shed.

As the reporting team follow the cows they see the farmer emerge from an outhouse and engage another farmhand in animated conversation. On this scene they record another 30 seconds of pictures before the farmer makes off in another direction, and the camera-operator stands watching through the viewfinder as the figure recedes into the distance.

Inside the shed milking is in progress. The camera-operator sets up the tripod close to where the action is taking place, the clatter of machinery and the lowing of the cows making effective background sounds. The light is low but acceptable, and the automatic exposure mechanism within the camera has no difficulty coping with everything except a wide shot of the whole interior. After he has recorded about a minute of pictures, the camera-operator is satisfied.

At this point, before unlocking the camera from the tripod, he rewinds the tape and reviews the previous two minutes of pictures through the viewfinder, putting on headphones to check the sound recorded over the shots of the herd on the move. Listening carefully, he confirms his suspicion that it was patchy and inadequate, and tells the reporter he is going back into the field to record some 'wild' sound of country ambience, using the long rifle-microphone supplied as part of his equipment.

The reporter, meanwhile, has found a quiet spot away from all the activity, and has begun to replay her own audio-cassette recording of the interview with the farmer. After hearing it once all the way through she decides that the second and third answers are easily the most relevant, summing up the issues most succinctly. Timed on her wrist stop-watch, they account for a total of 1 minute 25 seconds. Too long, she considers, for a package she was told would be allocated no more than three minutes on the screen.

Listening even more carefully to a second replay of the tape, she decides that by coming in later on the first of the chosen answers, the interview can be pared down still further without destroying the core of the farmer's argument, although one particularly colourful reference to government policy is lost. The interview is now down to 50 seconds, which is much better. Going back over the rest of the interview yet again, she makes notes about some of the other points raised, adding them to the references made during her original research. These are for possible use in the commentary she must now write to cover the rest of the pictures.

The reporter made sure to accompany the camera-operator throughout the entire process of picture-taking, and now she consults the detailed list she made as they went along. With the full knowledge of the content and duration of each shot, she can visualize how the edited pictures will eventually look on the screen when accompanied by the commentary. Without that essential information it would all have been down to guess-work, and would have shown as such.

The rough outline of the package is beginning to take shape as the reporter starts to weave her commentary around the pictures. Conscious of the need to identify the location as quickly as possible, she opens by setting the scene, trying to imagine the shots of the farm entrance, which will be taken as they leave:

> 'As farms go, Topfield is very much in the miniature class, not even one hundred acres.'

Next she wants to sketch in the background as briefly as possible, to convey an idea of the type of farming being carried out. There were plenty of shots of general activity immediately after the farmer's interview, and she rejects the phrase 'pig-rearing' as being too explicit, favouring the all-purpose 'stock-breeding' so the picture editor will be able to select the most suitable:

> '... but, thanks to a mixture of stock-breeding and milk production, every one of the past twenty years has produced a profit. Until this year, that is, when despite increased production and higher European subsidies, Topfield – like so many farms of similar size – is facing a loss. It's potentially so crippling that whatever the politicians in Brussels may decide today it won't be enough to save ...'

At this point the commentary begins to approach the central issue of the prize Jerseys and their future. The reporter remembers the series of shots taken by the camera-operator as the cows were being moved from the lower grazing field, and she sees the sequence extending naturally to the noisy scenes in the milking shed. Although continuity must be maintained, she deliberately avoids too rigid a construction which would leave the picture editor no scope to put in extra shots without having to make drastic alterations to the sound track:

> '... the herd of seventy prize Jersey cows from being sold off. Since the first of the strain was bought five years ago, the farm's average milk yield has more than doubled. Recent investment in a new, automated milking system seemed certain to increase production still further. Despite modernization, things have gone seriously wrong.'

The next paragraph leads into the interview with the farmer. For this the reporter expects the picture editor will want to use the two-shot, taken with the milk churns in full view. The last few words have to pose the question, phrasing it in a way which makes the chosen answer follow naturally:

> 'Topfield's owner, John Brown, has survived previous years of crisis. What's so different about this one?'

Including the 50 seconds allocated to the interview, the reporter estimates that, depending on the way the milking sequence is eventually edited, between 1 minute 45 seconds and two minutes have so far been accounted for. Although she might actually get away with another 20 seconds or so on top of that, she prides herself on turning in her items to length, besides which she knows the programme editor on duty today gets extremely annoyed with reporters who go beyond their brief without good reason. So a maximum of 1 minute 15 seconds is all that is left for the remaining ingredients, including the piece to camera, which took 32 seconds. The reporter makes the most of the obvious link between the farmer and the pictures of his wife, and thinks the farmhouse exteriors make a neater transition to Mrs Brown than going straight to her after the interview:

> 'The Browns bought Topfield when they married in the mid-eighties. At the time it was virtually run down, but Mrs Marjorie Brown, who used to spend weekends here helping out the previous owner, persuaded her husband it had potential.'

If necessary, the reporter reckons, the next line could be omitted to save space, although it would be a pity.

> 'She has a degree in farm management and had strong ideas about how to put things right financially.'

The final two sentences, about the future, are important. Here the reporter is in two minds whether they are better illustrated by shots of the farmhands and general activity, or by the conversation between farmer and helper recorded just before the milking sequence. Her careful choice of words means the office can decide:

> 'Now the Browns are considering that if the herd has to go they might as well sell up completely. If they're hesitating, it's because they're in no doubt about what it would mean for their tiny workforce – redundancy in an area where unemployment is already high.'

The commentary leading on from the interview adds up to 35 seconds. Adding on the piece to camera brings the total length to a maximum of three minutes seven seconds. 'Just right', thinks the reporter. If the office is desperate to save space, six seconds can be trimmed back by taking out the additional reference to the farmer's wife, and a few more by keeping a tight hold on the prize herd sequence. The whole scripting exercise has taken no more than 25 minutes.

Seeking out the camera-operator, who has now returned to the car, the reporter finds a place on the farm where no single sound is dominant, and reads her commentary, interspersed with editing instructions for the picture editor, into the microphone plugged into the back of the camera, the signals recorded onto the second of the two audio tracks on the videotape. She stumbles once, over the difficult phrase 'here helping out'. After a deliberate pause she substitutes 'working here for' for the offending words and she repeats the whole paragraph.

With the commentary now successfully recorded, the crew take their leave of the farmer, the camera-operator remembering to stop briefly for his final shots before they drive out of the farm gates. The whole operation has taken a little over three hours for a total of well under 20 minutes of tape – less than a single full cassette.

The team now face the prospect of getting their story back to meet their lunchtime news deadline, just over two and a half hours from now. As Topfield Farm is located approximately 80 miles from base, there is no prospect of driving the tape back through town traffic and getting it edited in time for the newscast.

Fortunately, sensible plans were laid long before they set out. A facilities engineer has already been detailed to rendezvous with them five miles away, at a point chosen for its height and relative proximity to the site of another story for later in the day. By the time the team arrive the engineer has already raised the aerial on his field car and aligned a test microwave signal with that of the nearest telecommunications tower, 20 miles away, and from there to a receiver at home base: it is a simple 'two-hop' operation which takes only a few minutes to accomplish.

The camera-operator, having rewound the tape, hands the cassette to the engineer, who inserts it into his own video replay machine. A brief check is made to ensure sound and vision signals are being received at the other end, where they are being recorded onto another machine in the main videotape area. The picture editor and writer from the newsroom are also there, watching anxiously and making notes about content and quality as the rushes are played over.

As the writer has a question for the reporter about one aspect of the script, the two have a short conversation about it over the link. The videotape operator has meanwhile made a swift spot-check to ensure the recording has 'taken', and then hands the cassette to the picture editor, who rushes off to his editing suite.

Using the same link the reporter calls the assignments desk, explains that her mission has been completed, adding that unless anything else has cropped up for either of them, she and the camera-operator will soon start making their separate ways back to base.

Editing the package

One of the features of modern editing is that it offers a variety of options. Editing a package is a fairly slow process as a rule, but excellent results are possible as long as reporters have played their part by taking detailed notes of the pictures and preparing scripts with them in mind. The best reporters are always prepared to adjust their scripts to take account of interesting picture sequences for which they might not have originally made allowance.

The main drawback of any rushes-by-link is that conversations conducted at a distance are not always as enlightening as they should be. An item which arrived directly back in the office would usually be accompanied by a minimum of paperwork: in some cases, particularly with overseas assignments sent home by air, a detailed script and shot-list. The use of portable and office-based newsroom computer systems has made an impact in this area, too, but this time, without the benefit of anything else, our 'Topfield' picture editor has to waste precious minutes in the edit suite listening to the commentary to get the proper 'feel' of the report, although he does have the notes he scribbled during transmission of the rushes, and the time-code on the bottom of the picture will be invaluable in helping him locate shots.

After another viewing of the pictures and listening to the interview with farmer Brown, the picture editor goes into 'fast forward' mode. He might be in an edit suite, or in front of a computer-based editing system inside a newsroom area. Wherever editing takes place attention must be given to the sound just as much as the vision. The editor reaches the time-code marking the beginning of the reporter's recorded sound. This includes a spoken guide to the way the edit should be carried out.

> 'This is the Delia Ward commentary for the Topfield Farm story for *The Five-O'Clock Report* for June the thirteenth. Beginning in five seconds from now.'
> *(Pause)*
> 'As farms go Topfield is very much in the miniature class – not even a hundred acres. But thanks to a mixture of stock-breeding and milk production, every one of the past twenty years has produced a profit. Until this year, that is, when despite

increased production and higher European subsidies, Topfield – like so many farms of a similar size – is facing a loss. It's potentially so crippling that whatever the politicians in Brussels may decide today it won't be enough to save the herd of seventy prize Jersey cows from being sold off. Since the first of the strain was bought five years ago, the farm's average milk yield has doubled. Recent investment in a new, automated milking system seemed certain to increase production still further. Despite modernization, things have gone seriously wrong. Topfield's owner, John Brown, has survived previous years of crisis. What's so different about this one?

'The interview with farmer Brown should come in at this point. I suggest you use the second answer beginning 'This time, we've not been able ...' which lasts about fifty seconds. The out words are 'whatever the government says'. Commentary continues:

'The Browns bought Topfield when they married in the early eighties. At the time it was virtually run down, but Mrs Marjorie Brown, who used to spend weekends here ... um ... er ... helping out ...

'Oops! Going again with that paragraph in five.' *(Pause)*

'The Browns bought Topfield when they married in the early eighties. At the time it was virtually run down, but Mrs Marjorie Brown, who used to spend weekends here working for the previous owner, persuaded her husband it had potential. She has a degree in farm management and had strong ideas about how to put things right financially. Now the Browns are considering that if the herd has to go they might as well sell up completely. If they're hesitating it's because they're in no doubt about what it would mean for their tiny workforce – redundancy in an area where unemployment is already high.

'The piece to camera comes in here. Use the second take. It lasts about thirty seconds. There's one obvious trim if needed. That's in the paragraph after the reference to Mrs Brown working for the previous owner.'

The picture editor starts his assembly by re-recording Delia Ward's scene-setting paragraphs onto one of the two audio tracks on the tape he has loaded into his recorder. Then he spins through the tape on the player until he reaches the time-code marking the matching shots of the farm entrance and general activity. He considers the camera-operator has done well to capture the air of bustle about the place, and wants to make sure that this is reflected in the final assembly. When the picture editor edits those scenes to the reporter's opening words he also records the natural sound effects onto the tape's remaining audio track at a level which ensures the commentary will still be heard comfortably above the noise. When the reporter pauses in the narrative the editor increases the volume of the background sound.

In this way the completed story will have the commentary recorded at maximum level on one sound track and the 'actuality' on the other, the level of this one rising and falling. When the item is transmitted the two sound tracks will be played simultaneously, reproducing the 'mix' at levels of volume the picture editor has created at his machine.

If he had opted to edit the pictures first and add the commentary later, the picture editor would not have been able to record the actuality sound with these delicate rises and falls because he would not have been able to judge where the gaps in the commentary were going to occur. This onus would have been on the reporter to shot-list the item after editing to make sure the words fitted the pictures, pausing where it was necessary to accommodate the actuality sound. A sound mix would still have been required, and this would have been achieved by putting the edited cassette into the playback machine and recording it onto another tape, mixing the tracks at the same time.

On the occasion of our fictitious Topfield Farm story, the picture editor's task has been made fairly easy because reporter and camera-operator have done a full, competent job – partly because they had the time to consider what they were going to do before they did it.

The main problem is that on viewing the rushes, the picture editor and the writer are not impressed with the milking scenes which were slightly underexposed and out of focus, and they would have preferred to leave them out, cutting the reporter's reference to the 'new, automated milking system', and keeping the next sentence: 'Despite this modernization, things have gone seriously wrong', which would have stood in its own right. But having already recorded the commentary onto his assembly tape and built up the first 40 seconds or so of the pictures, the picture editor is reluctant to start all over again, as any 'surgery' to remove the offending words would otherwise be impossible.

As the item also looks like being over length, they are therefore left with the option of trimming back the interview with farmer Brown or chopping the piece to camera by a few seconds: the picture editor prefers the second alternative, because it is simpler to execute, but the writer believes that because the reporter has chosen to construct the item in this way, losing the final words would make the report incomplete. As for the interview, this has already been pared to the limit and would scarcely be comprehensible if cut back still more. It is a dilemma the two have faced many times in the past, and they solve it now by listening again to the piece to camera and deciding that, at a pinch, they can lose the opening sentence without affecting the overall sense. That saves ten seconds, bringing the total duration of the package to 3 minutes 2 seconds – just about acceptable.

Writing the intro

Getting the story right in the first place and organizing the coverage so it can reach the screen with the minimum delay is not the limit of the reporter's responsibilities. Just as much effort is needed to ensure that even before it reaches the transmission stage the package is not made out of date, either by a commentary which fails to make allowance for the delay between gathering the material and its screening, or by ignoring any possibility that it might be overtaken by events. ENG has helped shrink that time gap, but the reporter still has to be aware of the possibility that events outside her control might yet take place.

For that reason Delia Ward's Topfield Farm script has deliberately made no more than a passing reference to the ministerial meeting in Brussels, even though this was the basis on which the report was originally envisaged. Her phrase, 'and whatever the politicians in Brussels may decide', will remain relevant until much later in the day, when she will be back at base and in a position to rework it if the report makes it into a later newscast. Delia Ward has also made the sensible assumption that an introduction based on the latest information from Brussels will be prepared by the writer for the programme presenter to read:

> 'The problems of European farmers are back on the agenda. Agriculture ministers meeting in Brussels are deciding today if small farms within the EU can be helped to survive in the current financial climate and at a time when production targets

have already been met. British farmers are not convinced that any action now will be effective enough to prevent bankruptcies among those who have barely kept their heads above water for years. Delia Ward has been finding out about the worries of one farmer in the heart of England.'

How the 'hard' facts of a news story are normally shared between reporter and studio presenter is a matter for item-by-item assessment, perhaps more than ever since journalist-anchors began writing more of the programmes they present. Nothing is more irritating to the viewer than to hear a report begin with exactly the same words or facts which the studio reader has used only seconds earlier. Without a formal system of 'sharing' the available information, which would be impractical, the prime objective is to ensure studio introduction and following report complement each other.

Although the temptation to take in all the details is at times practically overwhelming, the reporter has an inescapable obligation to the newsroom to leave at least some of them for the introduction. The choice open to those at the output end would otherwise rest between the evils of a bland, uninformative introduction (which no programme editor would willingly tolerate) and surgery, the carving out of some of the facts from the report itself for use in the introduction.

Fortunately, the need for such drastic action is usually unnecessary. After a sharp lesson or two, most reporters soon learn an acceptable level of self-discipline. Some achieve it by writing a full introduction of perhaps three sentences, and beginning the commentary proper with the third. The other two are then sent off with the script details or passed on in some other way to provide cue material for the presenter.

This admirably simple technique is capable of being used by any reporter in any situation. Yet it still manages to create difficulties for those whose imaginations are not as sharp as they might be. Deprived of the two most obvious opening sentences, some reporters are reduced to beginning the third in a form which, instead of helping the viewer pick up the thread formed by the introduction, has become a cliché:

'It/The explosion/accident ... happened ...'

Making 'films'

For all the modern refinements which have accompanied welcome improvements in technology and the expansion of continuous rolling TV news services, the average three- or four-minute news 'package' as the staple journalistic tool remained fundamentally the same through the late 1990s. A vague feeling of staleness has begun to emerge. The principle is not in doubt. What is in question is the somewhat formulaic approach to construction adopted by many reporters, together with an absence of what might best be described as 'filmic' qualities in packages made under no great pressure.

For sure, some stories require nothing other than straightforward, uncomplicated treatment, and no one wants journalists and camera crews to get their heads shot off indulging a predilection for fancy camera work in dangerous locations, but elsewhere reporters should be looking for the more imaginative pictures which will add something extra to otherwise standard subjects. A single sequence lightened by creative positioning of the camera – perhaps simply no more than an elevation above the action – and imaginative picture composition may be all that is necessary to transform a mundane package into a 'film'.

11

The camera at work

The way the camera is used and collects images has not changed in many years. The range of people who use it has changed completely. The reason is user-friendliness. Modern light digital cameras are simple to operate and have improved in their picture quality, although sound acquisition has become the biggest problem for any hopeful camera-operator. It is important to distinguish between using the camera for hard news and soft news. With the former you get one chance at the shot, and it can usually mean dangerous or demanding work with a tight deadline; the latter takes place in a more controlled and safer environment better suited to features or background material. There is a place in modern news for both ways of working.

Soft news camera

By 2000 the 'crew' in some programme units became an obsolete word, with the reporter working alone with a digital lightweight camera, bringing with it an operational flexibility and economy previously unimagined. On the other side of the coin there are plenty of professional camera-operators who took to the task of being roving reporters as well. The post-millennium news-gathering operation has now blurred the distinction between reporter and 'crew' in many smaller or narrow-focused programme areas. This also removes the distinction between amateur and professional and so we are familiar now with the term 'video-journalist'. In specialized programme-making teams working on softer features, such as feature film, music, lifestyle or travel stories, a person can have an idea, film it, script it, edit it on a computer screen and lay the track. It is fast, cheap, flexible and any hazards are easier to predict.

Hard news camera

In most of the mainstream news organizations there remains a combination of two skills deployed by two people – the one with the camera, and the one with the words. This remains especially true in covering 'hard' news events which are complex, dangerous, stressful, or all three. Many news events, by their nature, are still like that. That is why they *are* news. Despite the advances in lightweight technology, the case for the professional camera-operator for the tougher television news or current affairs assignment has remained strong.

Figure 11.1 Camera crews in action: ITN cameraman Nigel Thompson and reporter Paul Davies. Pressures on news services to make economies have led to a general reduction in crew numbers. Teams made up of camera-operator and sound recordist still exist but have largely given way to single-crewing. (Photo courtesy of ITN.)

Safety

A reporter and separate camera-operator can watch out for one another. The reporter working alone with the camera needs to have considerable regard for health and safety. With one eye fixed onto the camera eyepiece, walking backwards is not recommended, no matter how much the shot will be enriched. Riots and other forms of conflict are not suitable news events to be covered by reporter-video-journalists on their own. Video-journalists, the common alternative word for a combined reporter-camera-operator, can, however, get a lot out of feature events where the environment is easily controlled and the deadline is not fixed. For safety it still means paying attention to how and where the video-journalist stands: avoiding stairwells, precarious edges and behind doors which are likely to open without warning.

Camera equipment

Such is the pace of change and development of camera equipment that it is impossible to offer a definitive list of what an individual operator or video-journalist would be expected to carry. Much depends on the range of assignments they might be asked to undertake, personal preference for one type and make of lens over another, and their own working

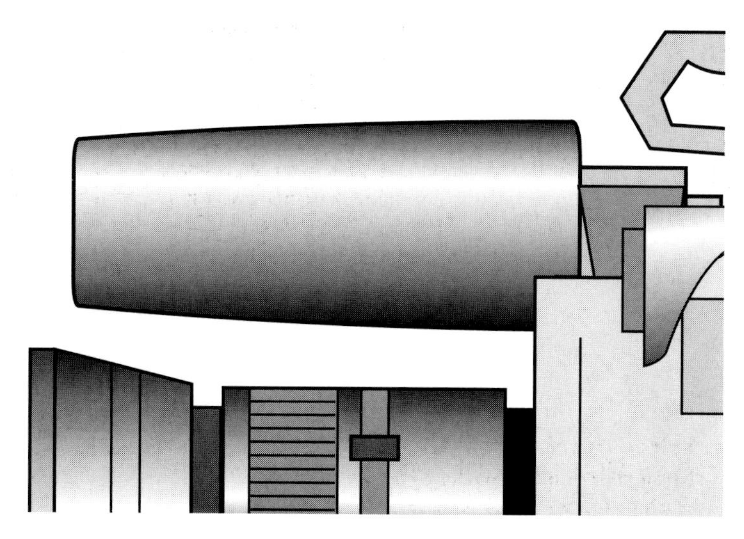

Figure 11.2 Camcorder built-in microphone. A useful device in many situations, but some occasions call for the use of separate microphones.

practices, developed perhaps over a number of years. However, what constitutes a basic set of either electronic or digital gear is unlikely to vary wildly from the following:

camcorder and/or camera and recorder
lenses
video cassettes (wrapped until needed)
batteries and charger (protected from humidity and climatic extremes)
tripod and mounting plate
mains power unit
basic lighting kit for more elegant shots (3 lamps)
stands and barn-doors
filters and diffusers
extension leads and connecting cables
electrical socket adapters
audio mixer
microphones
microphone stands

The batteries give enough power to drive the camera through about an hour of material, depending on conditions, before recharging is necessary. The mains power unit is used when the camera-operator knows that he or she will be able to plug it in to a suitable indoor location for a long period.

Microphones

Although cameras come equipped with built-in directional microphones, operators often prefer to use any one of a number of different types, according to location, weather conditions, and the availability of other help (usually the reporter).

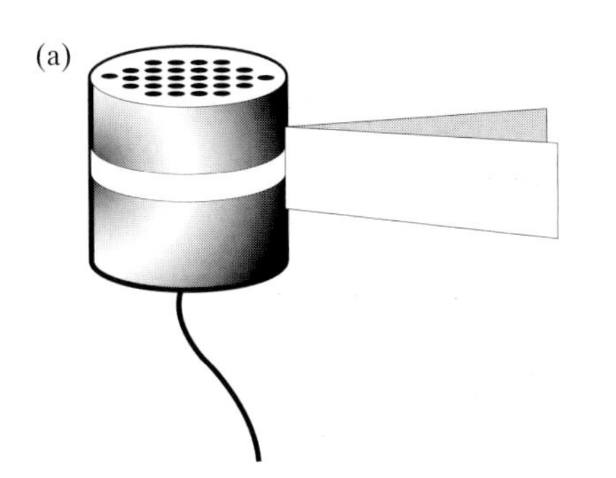

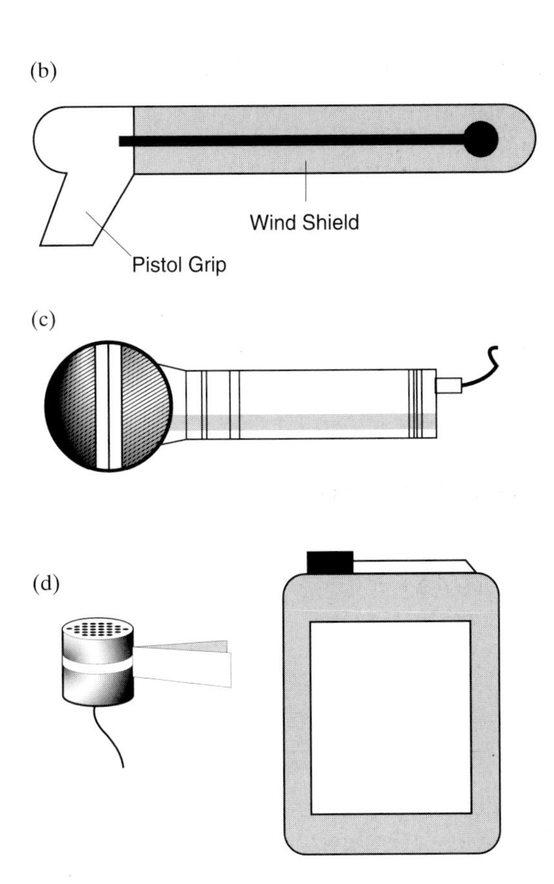

Figure 11.3 Microphone types.
(a) The battery-powered personal microphone clips to clothing at about chest height;
(b) the gun or directional microphone is usually covered in a light plastic tube;
(c) the simple stick microphone;
(d) radio microphone: a combination of a microphone and a smaller transmitter which can easily be hidden from the audience.

For interviews in particular the most favoured is the battery-powered personal microphone. This is small, light and clips to clothing at about chest height. As it is placed fairly near the mouth, this type of microphone is ideal for interviews or pieces to camera where background noise is intrusive. It is particularly useful when plugged into digital cameras.

The gun or directional microphone is usually covered in a light plastic tube as protection against wind noise. The gun microphone picks up sound through a narrow angle over long range, which enhances its versatility. But it is also inclined to restrict the camera-operator as the microphone (or whoever is holding it) can easily creep into shot by mistake. Its shape is also unfortunate: from some angles it closely resembles a weapon and in some situations this might draw dangerous attention to the user.

A third popular type is the simple stick microphone, held by the reporter. It has no handling noise, can be prepared for use in a very short time, and has the extra benefit of giving the nervous or inexperienced something useful to do with one hand. The stick microphone should be grasped firmly near the top by the fist, not truncheon-like until the knuckles go white or so delicately by the fingertips that it waves about out of control. If used for interviews it should be 'favoured' gently towards each speaker, preferably at a distance from each mouth so sound levels can be kept comfortably balanced.

Next comes the radio microphone, which has become one of the essential tools of news reporting. It has the distinct advantage of freeing the user from a lead attached to the camera – which troublemakers learned could be cut as a way of interrupting coverage. The piece of equipment comes in two parts: a microphone and a separate, pocket-sized transmitter. Some early radio microphones were inclined to be temperamental, but reliability has improved enormously and good reception can be expected over about 150 metres.

Other microphone types may be supplied as standard or made available from a general pool for special assignments. These include those fixed to booms above speakers' heads, and stand microphones, which are well-known for producing high-quality sound at the expense of obtrusiveness.

Camera-operators, being acquisitive creatures by nature, also manage to gather spare parts and various other bits and pieces of equipment over the years in an effort to make their working and personal lives on the road that bit more comfortable. Stout shoes, gumboots, warm clothing and a case containing passport, other documents, toiletries and underwear can usually be found in an odd corner in preparation for the unexpected foreign assignments. The issue of flak jackets, gas masks and other special protective clothing has also become almost a routine necessity in preparation for news-gathering.

Basics of camerawork

As with other areas of television news, there are few rigid rules which camera-operators must follow, so much depends upon the time, place and nature of assignment, and on the experience of the operator concerned. But there are some basic principles which the most conscientious have adopted and apply instinctively under almost all circumstances.

There is no substitute for a sharp, rock-steady picture wherever possible. Most of the time that will mean using a tripod, even though professional camcorders are ergonomically designed to be held comfortably on the shoulder. Sudden jerking or shaking of the picture is entirely compatible with the coverage of riots or similar situations, but such movements are out of place in set-piece interviews conducted at relative leisure. Small digital cameras have a problem as a by-product of their virtue – they are just too light. Their lack of mass makes it harder to obtain a steady shot if hand-held. In the absence of a tripod the video-journalist should try to prop himself or herself against the nearest firm structure (usually a wall, doorframe or even a tree).

Although fashions have changed with programme concepts, it is still acknowledged that news work should be kept direct and straightforward, without 'arty' movement for its own sake. Pans, zooms and other extravagances should be kept to a minimum, making sure that if they are used the beginning and end of each shot is held long enough for the picture editor to get the scissors in. Pans and zooms however are perfectly acceptable in longer current affairs programmes. In this case they are always thought out beforehand, and the camera movement will be done for a reason. 'Hosepiping', the rapid spraying of the camera in all directions, belongs to the media-wannabees who think their reports must look like one of those early music videos from the 1970s. Hosepiping works fine on Saturday morning live shows for the kids because 'live' means you can move like a human eye. At the other extreme the professional does try to avoid the boringly static scene which could just as easily be captured by the still camera. If a news programme is suited towards rapid movement of the camera, at least make sure it looks deliberate!

At all times the pictures must be shot with the editing in mind, which is why most training for combined video-journalists-editors must include sound and vision editing as well. That means avoiding single, isolated shots. Aim to produce sequences of long, medium and close shots, with plenty of variety in angles. Overdone, though, this can lead to complications. Picture editors and writers have been known to come almost to blows with camera crews who, they maintain, have swamped them with pictures they have simply been unable to view in the time available, and have therefore omitted key shots they were not aware existed.

International assignments

For world television news services with internationally minded audiences, the task of covering important events in far-off places can be daunting, fraught with uncertainty and at times almost crippling in expense. This means that while the less well-off are forced to rely on agency or pooled material picked up at relatively bargain-basement rates, the larger organizations which demand exclusivity and speed to enhance their international reputations have no choice but to join the rat-race, praying the budget will stand it and casting anxious glances over their shoulders to see how the competition is faring. This is despite the economic realism which governs foreign news-gathering as never before.

As a matter of course, any staff or freelance journalist who contributes regularly to a television news programme from a foreign base is expected to establish good local working arrangements and secure lines of communication (see pp. 21–22). Different criteria are applied to what is known as fire-brigade operations – those occasions when news teams are sent direct from home or the nearest suitable location to cover particular stories as long as they last.

Television news crews operate abroad under frequently untried conditions which vary according to country and circumstance. Local attitudes range from enthusiasm, with an offer of every facility, through tolerance to downright hostility and threats leading to restriction of movement, harassment or arrest. The depressing catalogue of death and injury which has befallen journalists in recent years, especially during the civil wars after the break-up of Yugoslavia in the last decade and the unrest in Indonesia in 1999, makes some accounts of earlier times seem almost relaxed by comparison.

Whether journalistic presence is a help or hindrance has also become a matter of controversy: well-fed reporters arriving to record harrowing scenes of starvation and departing again just as quickly; hampering humanitarian relief efforts simply by their presence; and contributing, perhaps, to the audience 'compassion fatigue' identified after a period in which the news is dominated by a succession of tales of mass human suffering. These charges are strongly rebutted by the news organizations, which point to the strong desire of relief agencies to heighten public awareness and which could not do so effectively without 'the media'.

During the conflicts which tore through Bosnia in the mid-1990s, the then British Foreign Secretary, Douglas Hurd, complained that some journalists behaved like 'founder members of the "Something must be done Club"', enthusiastically pushing for military intervention, in essence making political decisions more difficult.

For senior television journalists the answer was more straightforward. 'Television in particular is a very curious beast: it has huge muscles, distinctly poor eyesight and a disturbingly short attention span', said the BBC's foreign editor John Simpson. But what he suspected Mr Hurd disliked was the way television alerted people so graphically to what was going on in a place like Sarajevo because it made his job of edging away from involvement in Bosnia so much more difficult.[1]

Even where there is absence of controversy, where there is genuine readiness to cooperate, language and culture barriers, unfamiliarity of terrain and a whole list of major and minor differences provide their own natural obstacles in the path of efficient news-gathering. This does not include making any allowances for the solution to the other, perhaps more important half of the equation – how to get the material on the air at home within scheduled programme times.

Getting it home

Three options suggest themselves. The first and most obvious is for the crew to collect the material they have gathered and take it home at the end of the assignment. This may appear to be simple and foolproof, but 'hard' news has a notoriously short shelf-life, and the method is practicable only for the news-feature type of material, possibly collected on DV camera, where the time factor is not of overriding importance.

The second option, almost equally simple and obvious, is to send home, by air, each separate stage of the assignment as it is completed. Part of every camera team's routine at any foreign destination is a thorough check of all airline flight times for the quickest and most direct routes home. This 'shipping' of news material has become much simpler since airline staff employed in cargo departments started to become familiar with those precious 'onion' bags. Arrangements can be made for the entire cost of shipment except local taxes to be met at the receiving end. All it requires at most friendly airports is enough time (sometimes no more than half an hour before the flight) in which to conclude the formalities of a customs declaration and to obtain a form of receipt called an air waybill. To speed clearance and collection at destination the waybill number is telephoned, faxed or e-mailed to those at home.

Scheduled flights are not the only answer. Freighters and charters play their parts as well, and there are some occasions when the material is important enough for the news service to hire its own aircraft to fly the story out. This is very rare, but the big operators will still do it for an international scoop.

An alternative to shipping is hand-carrying it as the personal baggage of a member of the unit, air crew or cabin staff. On occasions it is entrusted to a willing passenger, sometimes referred to as a 'pigeon', who is met personally on arrival and gratefully relieved of their burden.

1. The Huw Weldon Lecture, 1993.

FOR CARDS
OUTSIDE
THE UK,
PLEASE
AFFIX A
POSTAGE STAMP

Ejemhen Esangbedo
Information Update Service
Butterworth-Heinemann
FREEPOST
Oxford
Oxon
OX2 8BR
UK

So to the third and fastest option, to which news organizations are turning in increasing numbers as the pressure mounts on them to provide instant or at the very least same-day coverage of events. Despite the cost, the use of satellites to send sound and pictures from news locations has become by far the biggest growth area in global news-gathering.

It was the use of this option which began to change the face of television news back in the late 1960s and, as many would have it, helped turn American public opinion against the war in Vietnam. Such a powerful phenomenon deserves close examination.

Telling it to the birds: the development of satellites in the late twentieth-century

Sitting apparently motionless 36 000 km (22 300 miles) above earth, man-made space stations are the agencies through which viewers of television news programmes the world over are able to witness the momentous events of their time – political changes, state occasions, natural disasters, sport, civil unrest and wars, of which Vietnam has gone down in history as television's first.

It was the steady drip of nightly newsfilm showing fighting in the jungle and the paddy fields which many believe finally sickened the American people into demanding an end to the carnage. Without the communications satellite to speed the coverage on its way between battlefront and living room, it is arguable whether the impact would have been as great, as soon. Vivid, full-colour pictures of this morning's fresh casualties have a gruesome reality that those of yesterday's do not.

Throwing television news pictures across continents was not, however, all that new. Even in its formative days, BBC TV News managed to receive material direct from the United States using a BBC system known as cable film. This employed conventional transatlantic sound circuits, but it was a slow process, requiring about an hour and a half for every minute of 16 mm film. An extra hazard was that the sound had to be sent separately, making synchronization another technical hurdle to be overcome before the whole could be transmitted.

The impetus for something faster and more reliable came, not surprisingly, from the United States, where potential military advantages reinforced a commitment to the exploration of space. The principle was to bounce the picture and sound signals off orbiting satellites from one earth station to another.

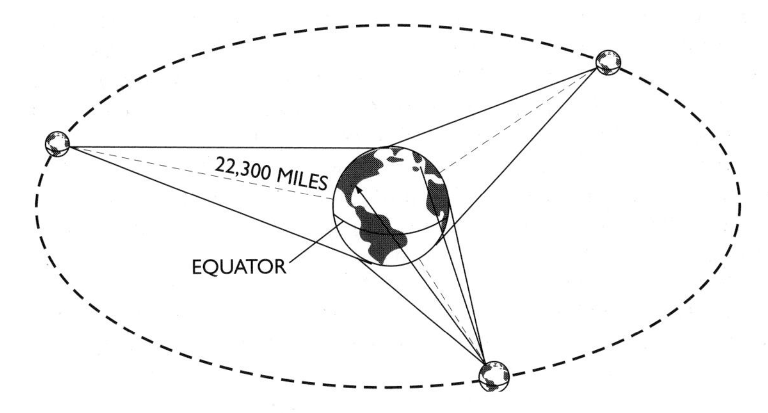

Figure 11.4 Geosynchronous orbit. Satellites appear stationary at 22 300 miles, enabling global coverage.

The first satellite was launched from Florida in July 1962. This was a tiny piece of electronic wizardry called Telstar. Every two and a half hours it completed an earth orbit, each one bringing it within range of ground stations built in America, Britain and France. The immediate effect was sensational: the only trouble was that the satellite was effective only for the few minutes that each orbit was visible to both sides of the Atlantic.

Within a very short time Telstar had proved to be only the forerunner of a world-wide satellite system for public use. This was the International Telecommunications Satellite Organization, more generally known as Intelsat, established in August 1964 with a founder membership of eleven countries which, by the turn of the century, had grown to 143. Intelsat operates as a wholesaler of satellite communications. It links the world's telecommunications networks. It owns and operates a global satellite system that links public networks, video services, private and business networks and Internet services. Each member pays towards operating costs, research and development in proportion to the use it makes of the system.

The United States is represented by Comsat, the communications satellite organization, and it is the biggest investor in Intelsat, with a holding of 20.2 per cent. Comsat provides telecommunications, broadcast and digital networking services between the United States and the rest of the world. These are used by Internet service providers, multinational corporations, telecommunications carriers, and United States and foreign

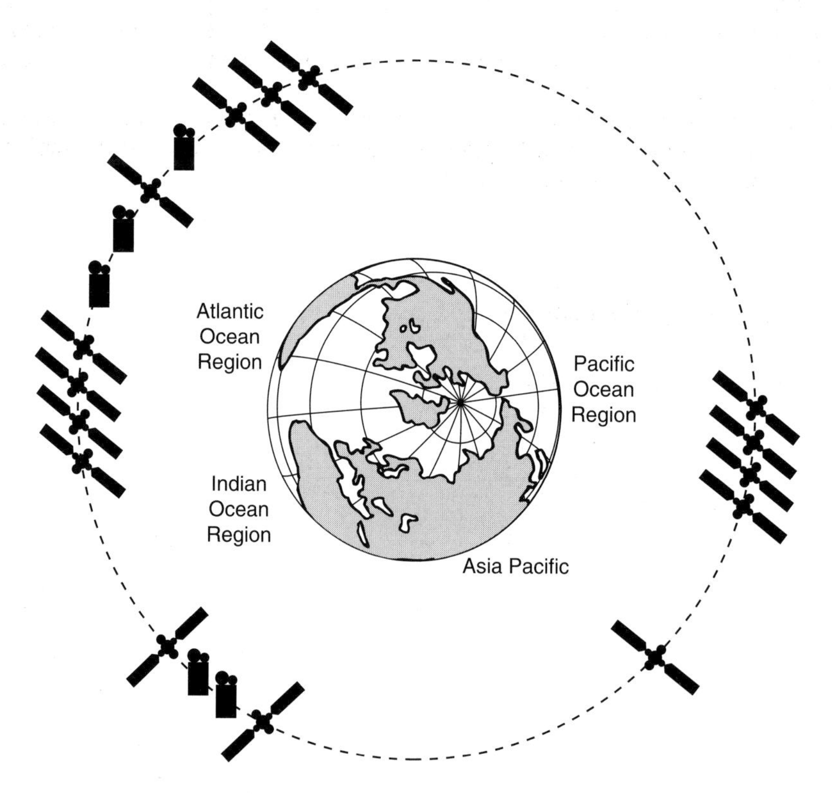

Figure 11.5 Intelsat has more than 140 members and operates as a wholesaler of satellite communications. The satellite system links public networks, video services, private and business networks and Internet services.

governments to extend their networks globally. The other main investors in Intelsat are the United Kingdom (8.3 per cent), India (5.2 per cent), Germany (3.4 per cent) and France (2.3 per cent).

The first satellite launched under the auspices of Intelsat was Early Bird, in 1965. Although it was capable of providing only one television channel for use between Europe and North America, this 39 kg satellite has won a permanent place in the vocabulary of television news journalists. To this day, satellites are known as 'birds', and 'birding' has become the accepted term for the entire process of transmitting news by satellite.

Early Bird, otherwise known as Intelsat I, was followed over the next five years by bigger and more powerful satellites. Intelsat II, placed over the Atlantic and Pacific in 1966–7, extended coverage to two-thirds of the world; the Intelsat III series completed the global link by the end of that decade. Further launches in the 1980s and 1990s ensured that broadcasters were able to transmit material from anywhere on earth using small, portable ground stations. By now each satellite appeared to be stationary in its appointed position over the equator – an idea first suggested by the author Arthur C. Clarke in a famous article published in the magazine *Wireless World* in 1945 – not only picking up and re-transmitting the signals, but amplifying them as well.

Until March 1984 all but two of the Intelsat satellites were launched from Florida, on the south coast of the United States, by NASA, the National Aeronautics and Space Administration. The European Space Agency's Ariane rocket system carried the others. Ariane has suffered its own setbacks, and the history of Intelsat has included some technical failures, with satellites lost when their launch vehicles went wrong and others not reaching their correct orbits.

By the end of 1994 the Intelsat chain had grown to twenty satellites, of which eleven were over the busy Atlantic Ocean Region, four each over the Indian and Pacific Ocean Regions and one covering the Asian Pacific.

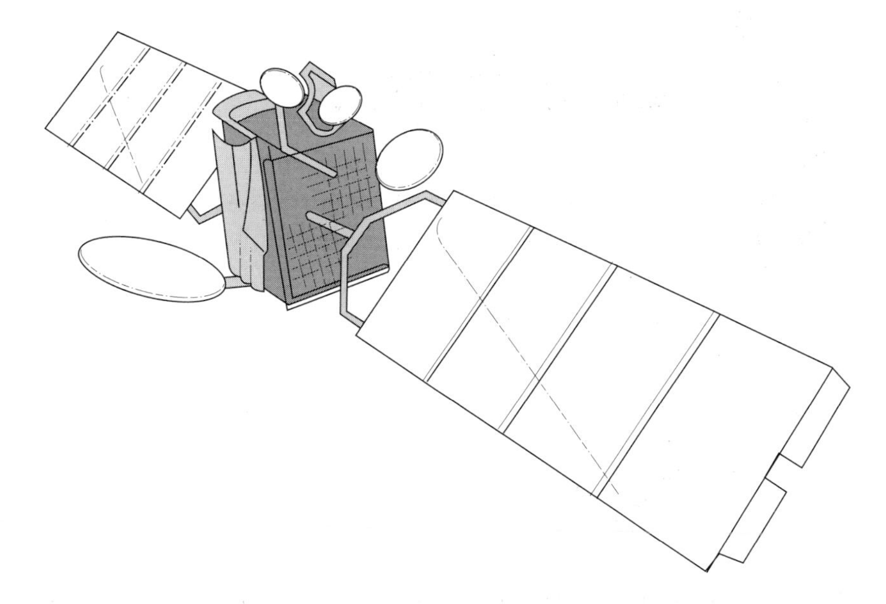

Figure 11.6 An impression of the Intelsat VIII satellite. Each will have a lifetime of between 15 and 18 years. (Intelsat)

A later development has seen the leasing of Intelsat capacity on a semi-permanent basis, especially between the United States and Europe. Reuters Television, the London-based international television news agency, joined forces with Western Union in BrightStar, a 24-hour direct video service using dedicated earth stations and microwave links in the United Kingdom and wholly owned earth stations in the USA, to link the transatlantic pathway with the American domestic satellite system. CNN International (CNNI) also took space on an Intelsat satellite to deliver its 24-hour news programmes from Atlanta, Georgia, to Europe and Latin America.

But it is the United States Information Agency which claims to have begun the first permanent global television network by satellite. Worldnet was inaugurated in 1983 as a means of getting across US foreign policy through live news conferences between American government officials and journalists in Europe and elsewhere. Two years later, a two-hour transmission service of news, current affairs, science and education features and sport was introduced, coming every weekday from USIA studios in Washington.

The success of the 'public' Intelsat system has also been followed by the arrival of competitors vying for a share of this increasingly lucrative business, among them the European-operated Eutelsat and Astra satellite systems. Those charged with the responsibility of bringing in by the fastest and cheapest route the raw material shot by their camera teams abroad, have a bewildering choice of carriers and tariffs, with frequent special offers bringing rates in off-peak periods down to only a few pounds a minute, compared with hundreds in the early days of satellite transmission.

Satellite news-gathering

The other enormous change in news-gathering operations has come with development of mobile satellite ground stations small enough to be packed away in a few boxes and carried into the field by air or road, making all but a few parts of the world geographically accessible. Signals from these 'uplinks' are aligned directly with a satellite in the global system, dispensing with the often time-consuming business of sending them to fixed ground stations.

The Gulf War, which engaged forces from the United Nations and Iraq in early 1991, proved it was possible to provide immediate, high-quality pictures and sound under even the most inhospitable conditions. Since then other front lines and big international events, such as the Solar eclipse in August 1999, and the celebrations for the Millennium, resulted in the growth of forests of satellite dishes operated by the world's leading television news services, and by specialist organizations working on their behalf, helping to increase the amount of coverage to a level never previously achieved. In turn the system has been successfully adapted for domestic use by news teams covering stories in remote areas of their own patch.

Digital technology, compressing the signals, has enabled equipment to become smaller, cheaper and more efficient, and development has reached a stage where it can be used to meet several communications requirements, including telephone, e-mail and computer network applications, and can be rapidly prepared for action by a single operator.

Meeting the deadline

For the television news team on a brief 'fire-brigade' foreign assignment, the availability and timing of satellite transmission is a crucial part of the planning of coverage, and

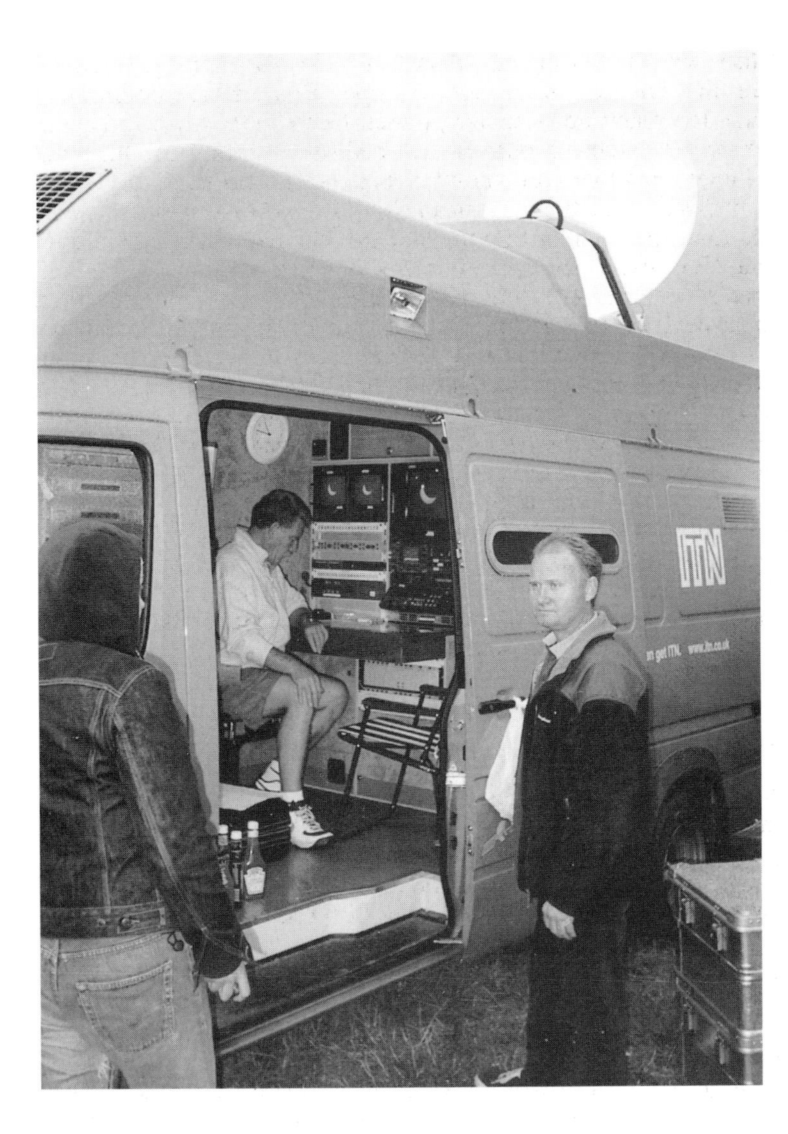

Figure 11.7 Satellite news-gathering equipment is getting ever smaller and lighter. The ease with which portable satellite uplink equipment can be transported to story locations has added a new dimension to news-gathering, whether abroad or in domestic sites. This ITN links vehicle is in Cornwall for the eclipse of the Sun. (Photo courtesy of ITN.)

the deadline these factors impose. Fine calculations on the spot are needed every time to ensure not only the news material itself is safely gathered (surprisingly, often the least difficult part of the exercise) but the pictures are returned from location in sufficient time to meet the demands of programmes thousands of miles and several time zones away.

Where really big running stories are concerned, the world's most important television news services fly their editing equipment to the nearest possible location and set it up in a hotel or similar centre. It means they are able to be independent of the local broadcasting organizations which might wish to be helpful but find themselves overwhelmed by foreign demands for facilities they themselves are scarcely able to use. In some places where a big news story breaks, hotels come to resemble broadcasting centres, with rooms converted into miniature studios, cables run under floors and satellite uplinks handily placed so that edited tapes can be fed direct. All this costs a great deal of money, because although conventional editing equipment is portable, it is not as portable as all that, and it is possible to visualize excess baggage charges continuing to have the effect of limiting foreign forays for all but the richest organizations.

Nevertheless, once the commitment is made and the bits and pieces have arrived, an experienced picture editor is able to turn almost any small room into a basic edit suite in under an hour. Even setting up and testing a modern PC-based laptop edit system takes as long. Certain compromises are sometimes necessary: since the surrounding ambient noise might be loud or unpredictable, the reporter might still use a traditional lip microphone (similar to those used by sports commentators) for recording commentaries direct onto the sound track, or a headphone pilot-mic style system. For the same reason, the picture editor might choose to use headphones instead of loudspeakers, though careful rearrangement of curtains and furniture in a hotel room can go a long way towards creating a passable imitation of studio conditions.

These conditions invariably help to generate a greater feeling of cameraderie and close cooperation between camera-operator, reporter and picture editor than is usually considered possible at home, because they are all working on the same story for a considerable time. The picture editor may be able to see the material gathered and suggest how it might be shot to help at the editing stage later, a distinct advance over most other stories to which he or she probably comes cold. The camera-operator may be able to spend some time watching the edit and explaining how a particular shot or sequence was constructed; and the reporter may be able to work at the same time as the picture editor, recording commentary during the assembly stage, changing it perhaps only minutely to extract the best possible combination of pictures and sound.

When the moment comes to transmit that precious edited package the team will do so either by feeding it back to base by way of their own satellite dish or, perhaps, taking it off to the local television station for onward transmission. All of course is based on the assumption that the necessary arrangements have been made in advance. There is no point in booking expensive time on any satellite unless some sort of prior arrangement has been made with those expected to provide the necessary facilities. There is no point in turning up in great haste at a heavily guarded television station without some form of acceptable documentation or at least the correct name of the current local coordinator or contact. Trying to explain to a soldier with an itchy finger on a gun that it has all been arranged 'by the office' and would he please stand aside to let you and a large hire-car inside at once is not the most enviable task to be faced with in a strange country. The chances are that you are there at a time of local tension anyway. The arrangements may indeed have been made 'by the office', but the office is a good 3000 miles away, and what may have seemed like a firm promise made to them at high level over the telex yesterday turns out to be a commitment to provide every facility tomorrow. Today is a national holiday you did not know about and only a skeleton staff is on duty.

Most of all, there is no point at all in arriving with edited stories or a mountain of cassettes of the most exciting news pictures recorded on a technically incompatible standard which cannot be transmitted or converted. So if disappointment or outright failure is to be avoided, it is essential for someone to have done all the necessary homework before the team ever set out. It might result in a decision to set up a separate facility known as a feedpack, a mobile standards converter through which it is possible to transmit material back to base. But that means another body, another hotel room and more excess baggage to add to the bill.

Often a better way out is to work on the local standard, be it PAL, NTSC or SECAM, and at the same time take advantage of a phenomenon of the world circus in news, which allows both formal and informal interchange of help and – often – raw material.

No longer do teams work in isolation. Now the international brigade is part of a great baggage train moving from one major event to another. No longer do crews shoot exclusively for themselves on the bigger stories. Coverage is carefully coordinated in partnership with others, even the usual competition, so a wide range of events and locations can be covered in a short time by a minimum number of crews.

The downside is the pressure faced by crews in the field. Bill Nicol, a BBC veteran cameraman whose experience straddles news-gathering using both film and video, defined the problem this way:

> Gone are the days when one could start early in the morning, do the story by lunchtime, put the film on the afternoon plane and then have a few hours to yourself. Now one still starts early and tries to complete a story in the morning so that the on-site picture editor will have plenty of time to assemble the package for satellite transmission into the first programme at home. Then it's back on the streets or up into the mountains to find the pictures to update the package for the next programme.

The demands of television news programming, intensified by continuous news services in both radio and television, often mean that reporters spend so much effort 'filing' for different deadlines they have little time and energy left to gather the news they are expected to transmit. Tales exist of correspondents who are so chained to the satellite link they have had to rely on news agency material fed to them from home about events they are meant to be covering.

The field producer/fixer

An increasingly important person in the international baggage train of the big story is a somewhat indeterminate figure known as a field producer or, more aptly, fixer. Crude though this latter title may be, it does accurately convey the essence of the task, as a member of the visiting news team to tie up the many loose ends at the scene of a major event, so that transmission into bulletins at home may proceed with the minimum of delay.

The fixer's duties are predicated on the fact that the reporter and camera-operator covering any news story are capable of being in only one place at a time, either in the thick of the action on location or back at the hotel editing suite, uplink site or local television station, taking up valuable time to negotiate such tiny but crucial matters as the allocation of space to keep the equipment safely and the arrangements to supply food and other necessities. Because the fixer is there, the rest of the team can throw themselves into their assignment, secure in the knowledge that wherever their temporary headquarters may be, one of their own is protecting their interests, keeping an eye on what the competition is doing, watching for any unexpected developments which might affect the story or their own circumstances, and soothing the anxieties of programme editors and other back home by providing frequent progress reports by telephone or e-mail.

The fixer therefore needs to be trusted for a sound editorial judgement, and as the possessor of a sound enough technical knowledge to be able to cut corners in a crisis.

The work is sometimes extremely boring – sitting about for hours in a foreign television station far from where important things are happening; frequently testing – trying to persuade an uninterested local technician to attend to your needs before those of your national competition; but almost always worthwhile. Assignments, often given on a temporary, story-by-story basis, are much sought after, particularly when they involve foreign travel, and the glamour scarcely seems to pall even though the most important task of an entire mission might be no more creative than jumping into a taxi to find a spare part for a camera that neither the normally resourceful camera-operator nor anyone else is able to produce.

Fixing roles, once invariably the province of experienced journalists allowed to escape briefly from newsroom-based duties, have also come to be assigned to picture editors, on the not unreasonable grounds of the cost of the additional person.

Those who have travelled the world as fixers will readily recognize the occasions when editing equipment has broken down and they have either had to repair it themselves or summon expert help; when in the absence of everyone else a snap decision has had to be made about the cancellation of satellite booking time, or the agreement to cover events which go beyond the original brief; when outside attempts to influence the editing politically have had to be resisted; and when it has been necessary to break a legal speed limit in an unfamiliar vehicle on the 'wrong' side of the road in order to get to the feed point in time. Add to that the fatigue which comes from never really being off duty away from base, and it is easy to see that, despite all the superficial attractions, fixing abroad is a task requiring real stamina and dedication.

Even on those occasions when all does go according to plan, reporter and fixer find the time towards transmission appearing to melt away at a disconcerting rate, until the realization comes that only a few minutes remain and the last edits have yet to be made. For the reporter and crew, on whom the news team at home are depending so much, the tension becomes almost unbearable. Will that effort be wasted, after all, in the dreadful anti-climax of a missed deadline? Fortunately, persistence and a refusal to panic have a tendency to pay off.

12

Constructing a news programme

Style and fashion change in television news just as much as they do in any other walk of life. The news programme today bears little resemblance to the programme of only ten years ago. Content and emphasis change. Today, that means more news about the reasons for crime rather than stories about crimes; more news about how children and adults learn than just stories about education in general; more about the environment and information technology and what they mean for the future; more analysis about the consumption of some nations and the poverty of others. News has always been tied up in the process of change. Old orders fall and, with that, new uncertainties arise. Television is therefore put together by people who are interested in what other people both want and need.'It means, fundamentally, that the people making decisions about news are interested in what a substantial section of the population is interested in at *a particular moment of time*.'

The audience of the new millennium is far more discerning, with the same high standards of presentation and production demanded of the news as of any programme which might have taken months rather than hours to prepare. So because television news cannot appear to be lagging behind in professional gloss or technical excellence, or appear to be divorced from media such as the Internet, the journalists have been drawn by the need to produce not only news on television, but news *for* television.

The news 'programme' is born

Although short bulletins and summaries still exist on many terrestrial and cable services, in the main the list of unrelated events has gone, to be replaced by programmes on terrestrial and cable thoughtfully constructed and prettily packaged. Periodic 'relaunches' also take place, often coinciding with changes at the top: designers are brought in to sweat over the shape, size and colour of studio sets, famous contemporary composers commissioned to write a few bars of stirring title music. Relaunching can be a seriously expensive business, especially if it involves complicated alterations or building work to move a programme's base from the newsroom to a new studio, or vice versa. A lot of serious investment goes into colour schemes. The big news organizations around the world tend to swing every few years between minimalist 'trust-me-I'm-serious' cool colours (blues and greys) and 'trust-me-I'm-friendly' softer pastels and natural shades (strawberries, cream and wood).

As part of a new look, old faces are likely to disappear from the screen and new ones take over, often amid much publicity and speculation about their salaries. The public has come to expect an element of showbiz about the news, even if it is merely confined to the window dressing at the end, when the presenters visibly relax, allow fleeting smiles to cross their previously grave faces, finger their scripts, and exchange pleasantries with other occupants of the studio. (One of the questions most frequently asked about double-headed presentation is what on-screen partners say to each other during the final few seconds they are on the air together.)

The importance of these trimmings should not be underestimated. Just as a newspaper properly seeks to attract its readers with the layout of its pages and its typographical styles as much as with the quality of its content, so the television news programme has to find a way of capturing audience interest and holding it right through to the end of its allotted time. Now more than ever, audiences are being asked to grasp abstract and complex issues which have a direct bearing on their lives. They have no chance of comprehending even a small percentage of them unless the subjects are presented clearly and unambiguously.

There is, however, one overriding factor: duration. Television news comes in all shapes and sizes ranging from summaries lasting a minute or two to marathon feasts of an hour or more, or continuous news services which are completely open-ended. Duration is the fundamental influence on style. The shorter the programme the shorter the items within it, the less room for frills, with only the bare bones of the day's news capable of being squeezed in. The longer the programme the greater the opportunity to spend time on explaining the issues, on casting the net more widely, and on employing the full panoply of television techniques.

But how long is a long programme? Peak-time half-hour news programming was the pattern in many parts of the world since it was pioneered in the United States, but breakfast and daytime programmes in particular have room to take things at a more leisurely pace, with shortish bursts of straight news finding a place between feature-type material, sport and weather. 'All-news' channels, like BBC News 24 and CNN, may appear to flow in a spontaneous way, but even here the format is based on segments or strands, and schedules are cleared to make way for continuous live coverage only on breaking into a segment format.

The argument in Britain many years ago was over what was called 'a bias against understanding', a phrase which summed up the view, that, although most events formed part of a continuing process and could not be dealt with in isolation, television news,

> devoting two minutes on successive nights to the latest unemployment figures, or the state of the stock market, with no time to put the story in context, gives the viewer no sense of how many of these problems relate to each other. It is more likely to leave him confused and uneasy.[1]

Coincidentally or not, the emphasis did begin to change, long before the holder of those views, John Birt, then working for London Weekend Television, found himself in a position to influence the journalistic approach within the BBC, where he became first Deputy Director General, then Director General from 1992 to 1999. In-depth reporting, analysis, background – call it what you will – began to find its way onto the screen in a form which in the past would more normally have been reserved for the longer, current-affairs type programmes. It became quite usual for news reporters and

1. *The Times*, 28 February 1975.

specialist correspondents to devote several days to the assembly of a single item intended to explain or interpret news as well as simply report the facts of it. The old idea of the 'bulletin of record', with every item of importance given an airing, however brief, largely disappeared. Instead, editors were encouraged to be more selective, to recognize that because time is so precious in television news it is better to tell fewer stories in greater detail. The change of emphasis heralded the introduction of longer and more frequent news programmes, the flexibility to overrun normal durations and, rarely, to interrupt schedules when very important events occurred.

All this assumes everybody actually wants more news on television. There are those who would say there is already too much, and occasional attempts on the part of those within the industry to reduce the power of the news organizations become a matter for national debate. A political outcry followed when the independent television companies removed ITN's *News at Ten* from the schedules, the aim being to clear peak evening viewing time for programming considered to be more commercially attractive. The ITN late evening replacement – a bulletin at 2300 – had lower ratings after the change.

Grumbles about news are that it is too expensive, takes up a disproportionate share of facilities or is placed so awkwardly in the schedules that the really popular, audience-pulling, advertisement-generating shows and films are losing out. As more channels become available and competition from other media fiercer it is possible to imagine the argument for removing news altogether from conventional schedules becoming compelling.

Most journalists, while happy to accept the arrival of 24-hour news as a worthwhile extra, would react with horror to the idea of being ghettoized. While every serious-minded news practitioner is always eager to put forward sound reasons for more programmes, together with enough extra resources to support them, the honest ones are privately prepared to admit that, for the time being at least, news is only one tenant in the crowded world of television programming. Drama, films, light entertainment, education, sport, information websites and the rest have the right to live in it too, and it would be entirely wrong of journalists to ignore the truth, however much it hurts: they have to fit in with the schedules and not the other way round.

Consolation lies in the knowledge that television news does not operate in a vacuum. The journalist who genuinely believes in the importance of communicating current events to the widest possible audience will rejoice that others are sharing the burden. Although it may well have been proved beyond doubt that more people obtain their information from television than from any other source, it is equally true that television is not and cannot be the sole provider. Radio, newspapers, magazines and the Internet all have their own contribution to make towards the sum total of knowledge, and it would be foolish as well as arrogant for television newspeople to think otherwise.

The onus on all those engaged on news programme-making is therefore to make certain they use every second of their available airtime to present the news attractively as well as intelligently. And that demands, more than anything else, recognition that the interests of the audience are paramount. That may be saying the obvious, but it is a principle which needs to be restated and re-emphasized, for in the technological revolution which has all but overwhelmed the journalists during the past few years, the basics are in danger of being overlooked and the message distorted.

Put at its simplest, it is all too easy for the professionals to assume that because they understand what they are transmitting the viewing public will do the same, taking for granted everybody's equal ability to concentrate on the news for as long as it lasts while viewing under ideal conditions. Even if this were the case (and it is possible to think of a hundred reasons why it is not) it is inevitable that levels of comprehension differ from person to person. Experiments to test the ability of viewers to recall in detail programmes witnessed only a short time earlier indicate concentration spans fluctuate and attention wanders during the course of a single bulletin.

Logic suggests this ought to apply less to programmes targeted at specific socio-economic groups, based on the outcome of comprehensive market research into style, content and on-screen personalities, but if, for whatever reason, some viewers do not take in what is being aimed at them, a proportion of those expensively gathered satellite pictures or the golden words of highly-paid presenters is going to be wasted. The best the practitioners of news can hope to do if they are to succeed is to reduce the comprehension fall-out by keeping their programmes and the items within them direct and uncomplicated.

Stories which are well-chosen in the first place, then carefully written and edited to present the most important facts with clarity and simplicity, are perfectly capable of conveying the message, even in a short time. Not so the irrelevancies and repetitions which masquerade as depth but add only length. Swiftly moving picture sequences, minuscule soundbites woven in and out of complicated packages, jazzy, all-action headlines and beautifully crafted graphics packed with information are all very well in their way. However much they may impress the boss or television journalistic colleagues, if there is any likelihood they will be lost on the viewer, what's the point?

No one pretends it is easy. The frenetic excitement generated by a busy news day can militate against cool judgement, and the temptation to use every marvellous electronic toy now at the disposal of television journalists can be irresistible. At times producers and editors succumb too easily to the temptation of transmitting live pictures from a scene just because it is possible to do so using mobile satellite or microwave links, without thinking whether what is seen and heard will enhance the product. The extra ingredient is frequently no more compelling than a reporter on the spot virtually repeating the contents of the account he or she has just transmitted. 'Live doesn't always mean lively', is how one senior television executive puts it. So why do it? For cosmetic reasons in some cases. In others, cynical acceptance that (a) 'lives' are cheaper because they use fewer resources than conventional filmed packages, and (b) if the links are not used often enough in this financial year the burgeoning accounts department will target them in the inevitable cost-cutting exercise before the next.

When rushes are flooding in by satellite, that most precious commodity, thinking time, can be in short supply or non-existent. The 40 minutes or so it once took old celluloid film to pass from processing bath to cutting room used to give programme editors a breathing space in which to make a considered assessment of whether they would actually want to transmit the stuff when it was ready. Beyond 2000 the wonder of instant video can dazzle an editor for choice, and it takes a strong mind and a refined news judgement to decide to reject late material which others have gone to considerable lengths to secure.

Putting it together

In the past, television news was so short, that editors preferred not to waste even a few seconds in reciting the contents. The items followed soon enough anyway, crowding one upon another, unannounced, at breakneck speed, more or less in order of importance.

If bulletins looked like overrunning, cuts could be made from the bottom up without seeming to disturb such overall shape as there was.

Since airtime began to be more generous, this philosophy has been made to seem out of date. In its place has evolved the concept of the television news programme, dependent for its success on the ability of those in charge to take a series of disparate events and fashion them into something capable of taking on a recognizable identity of its own.

The criticism is still sometimes made that in reaching out for that goal, editors are inclined to let themselves be over-influenced by the availability of pictures. Such a generalization seems impossible to prove one way or the other. Yet, if it is true, there seems little shame in admitting it. By what other criterion should a medium which deals in pictures base its judgement? Given a reasonable alternative, no editor would choose to open a peak-time television news programme with an indigestible wad of copy-only stories and studio reports, leaving the first pictures until ten minutes have passed. It should never mean ignoring the important non-visual story in favour of the trivial pictorial one. What it does mean is encouraging editors to apply to television news the values of television, as opposed to those of newspapers. Most of the time broadsheet front pages and television news will follow similar lines: when they do not, editors should beg to differ and go all out to exploit the advantage they have over the printed word.

In many ways the argument is not so much about what constitutes a good story as about emphasis. On a front page of a newspaper, clever layout is used to direct the reader's eye quickly to the most important item, or to any one of a number given equal prominence. In television news the implication is that order of importance is synonymous with the order in which events are transmitted.

In reality, whether or not they always succeed, some television news editors would prefer to concentrate on making programmes which viewers find easy to follow. Instead of being sprinkled haphazardly throughout the news like confetti, stories are sorted into small groups. An item about domestic industrial output, for example, might lead logically to one about exports, which puts the audience in a receptive frame of mind for a report from abroad. Brick-by-brick, the programme edifice should be built up in this way: little sequences of events linked by association of subject, geography or both.

Individual story durations and treatments have to be considered in parallel, so successive items will not look the same. This may lead to some stories being detached from one group to join another, or made to stand in isolation. There is no virtue in constructing a tortuous link for its own sake, or in promoting a story far beyond its importance just because it seems to fit. Without making a fetish of it, the target should be to produce a programme which has a beginning, a middle and an end, and which looks as though some thought and care has been given to its construction.

An essential part of the formula is the menu or headline sequence, which usually forms part of the titles. The headlines are meant to summarize each outstanding item in the programme, often in no more than a single sentence aimed at grabbing the attention of the audience and keeping it to the end. Over the years, headlines have developed from being straightforward reads on or off camera into proper sequences in their own right, with combinations of stills, graphics and videotape extracts to whet the appetite for what follows.

As well as imparting urgency at the top of the news, the headline technique also introduces a useful degree of flexibility. Once the bald details have been given, there is no rule to say the full reports themselves must follow each other in a block or, indeed, in the same order. Instead, editors should welcome the freedom to distribute their 'goodies' at points which help to give their programmes pace, variety and balance.

At the same time, other options have presented themselves for editors and producers eager for some format which will make their programmes stand out. Ideas are tried and discarded, varied and tried again.

Scores of details need to be settled. Here is a small sample:

- whether a programme should be fronted by one, two, or more presenters and in what gender, age and ethnic combination;
- who is regarded as the senior in a multi-presenter format, and how their working pattern should be planned;
- whether the lead presenter should preface the opening item with 'good morning' or 'good evening' or plunge straight into the news;
- whether studio performers should be framed dead-centre or offset to one side of the screen, and in medium shot or close-up;
- whether presenters should be seen also as reporters on location, and if so what happens in their absence;
- whether any separate sports and weather presenters should be included in the opening shot;
- whether studio introductions should be read by one person, off-camera commentaries by another;
- what kind of colour of background should be used: whether a plain, pastel studio wall, some illustration to suit each story or a generic programme symbol;
- whether correspondents or reporters appearing in the studio should be set against the same backgrounds;
- what 'house style' should be established in maps and other graphics;
- whether each contributor should be introduced verbally or by a name superimposition: if the latter, where it should be positioned on the screen;
- whether a 'brand name' or clock should be displayed at all times and, if so, in what size and position;
- whether sport should be included automatically or on merit;
- how a news programme should treat items immediately before and after any commercial break;
- whether the latter part of a newscast should contain more feature-type material;
- whether, in the absence of a commercial break, some other halfway sequence should be constructed;
- whether opening and closing title sequences should be accompanied by music or sound effects;
- whether there should even be set titles instead of something which changes daily according to programme content.

All this and more has to be decided with as much care and consideration as the way the news itself is reported, for without being offered high production values, the audience may not be inclined to keep watching.

How far editors are prepared to go in attracting and keeping an audience is another matter. High viewing figures – and the lucrative advertising which accompany them – can probably be achieved on a largely down-market diet of chilling crime and human interest stories fronted exclusively by good-looking journalists. Nevertheless, modern market research techniques are useful tools to be employed in establishing audience preferences, and most news organizations feel a need to bring in the style-doctors and focus-groups to help them relaunch news programmes.

Preparing for action

An 'act of faith' is how a senior BBC colleague once described the editorial, production and technical processes connected with the construction of every television news programme: faith in the certainty that each separate member of the news team is carrying out his or her allotted task while others are doing the same. It is a faith shared confidently with those in the field at home or abroad, in the newsroom, in the editing areas, in the transmission areas, in the studios and everywhere else associated with the apparatus of news. At the root of it all is good communication, without which, in the constantly shifting sands of news, the whole thing would probably sink without trace.

To the uninitiated, the daily home and foreign news diaries, the 'prospects' setting out detail of coverage, the wall-charts and computer screens on which the progress of every assignment is followed, probably seem confusing and unnecessary. To those engaged in the serious business of making something tangible out of a set of elusive hopes, promises and expectations, sometimes on the basis of events yet to happen in places thousands of miles away, they represent a comforting reminder of the amount of effort being expanded for the sake of a common goal.

Programme meetings

Although the variety in size and importance of individual organizations involved in daily television news makes it impossible to identify a single example as an illustration, it would be fairly unusual to find an office working pattern which did not make allowance for at least one morning editorial meeting to ensure as many people as possible are aware of plans likely to involve them at some stage over the next few hours. The idea is to provide a solid impetus to the team effort required, to enthuse everyone likely to be involved in the day's production, and the most successful meetings, whether held in a newsroom or conference area, are those open to as many people as possible, and not confined to the editorial staff. In some quarters 'pre-meeting meetings', attended by only a small group of executives, have become part of the routine. The suspicion here – justified or not – may be that this time is used to determine the shape of the day in a way which pays lip service to what goes on at the later gathering. And the really paranoid can only be left to guess whether 'pre-meeting meeting meetings', restricted to a cast of the most senior or favoured few, cover more than a discussion of the day's events.

The timing of morning meetings is also important. They should be held early enough in the day so anyone not already engaged on an assignment is able to attend, but should not be allowed to drag on or be side-tracked into irrelevancies. No more than half an hour of brisk, businesslike discussion should be sufficient for all but the most complex programmes. The crowded, tobacco-smoke-laden morning meetings I attended on a short but concentrated advisory mission to a newly liberalized national news organization in Eastern Europe tended to degenerate into long-winded arguments about journalistic freedom and integrity to such an extent that reporters and crews were invariably late getting out on the road and occasionally were too late for their assignments. While it was easy to empathize with those who feel a need to discuss such matters, the management was at fault for not keeping the meeting properly focused.

The significance of morning meetings in the culture of any organization can usually be measured by the seniority of those regularly chairing it. If it is the executive in charge of news – not a programme editor – who presides, then the occasion will be given a sense of importance and gravity which otherwise might be lacking. In the case of national programmes it is here that the duty editors responsible for home and foreign assignments will set out what is on their agenda for the day and how items in their domain are being handled, and they may also take a few moments to give first details of newly breaking stories or of developments in others.

It is here, too, that anyone engaged in the process should be encouraged to put forward ideas for coverage or treatment. Many a piece of original journalism has started with a half-formed idea which has been tossed backwards and forwards until it emerges as a rounded plan to be acted upon. Editors are known to complain that not nearly enough ideas are forthcoming from 'the troops', but it is often because those who would like to offer ideas believe the better part of their programme has been predetermined to a point where room is likely to be found for only the hardest of hard news. They are also likely to feel uncomfortable about putting their thoughts forward in company, for fear they may be ridiculed by senior colleagues or their peers. Ideas should always be encouraged and put-downs of unsuitable ones undertaken with great sensitivity, on the basis that a brusque rejection might deter another idea which would have hit exactly the right spot. Equally unproductive can be the routine in which the boss does the roving finger job – demands a contribution from everyone in turn. Not good management and not a way to raise morale; a sure way of encouraging people to find excuses for non-attendance.

For trainees and newcomers, editorial meetings can present something of a dilemma. Pipe up too often or too soon and your immediate colleagues will mark you down as pushy and arrogant: keep quiet and the editors and producers you hope to impress will think you have nothing to offer. Unless you are asked directly to contribute, your best plan is to hold back for a week or two to observe the rituals before risking a first foray.

Post-mortems

Much the same advice goes for post-mortems, which may or may not be part of a news organization's culture. Some editors from the Macho School of Management like to gather their teams around them immediately after programme transmission and perform a loud, public and searching review of successes and failures. Especially failures. Others wait until next day for a discussion, constructive or otherwise, which becomes part of the morning meeting.

In the knowledge that those who have usually poured their best efforts into a programme cannot wait to get home or to the bar, my own preference has always been for a very quick and low-key post-programme assessment, which is sure to include a few well-earned 'thank yous' and 'well dones'. More negative aspects are tackled in private. Professionals know when they have made mistakes or underperformed: humiliating them in company seems pointless, especially as it can soon undermine confidence, which in turn leads to more mistakes. Of course, no organization is perfect and it is always possible to discover new ways of getting things wrong. Since there is usually nothing to be done after the event about such irritations as a misspelled place name on a map, an example of poor writing or a badly constructed picture sequence, a 'discussion' which degenerates into a slanging match is a waste of time, especially when it will probably be necessary to start the process all over again tomorrow with the same people, and the very rare occasions on which I allowed myself to stray from this principle have always been a matter of personal regret.

The running order

In addition to morning meetings embracing an entire newsroom staff, many organizations have formal or informal gatherings for individual programme teams. While editors and producers will be sure to keep track of progress during their duty day and the good ones certainly never lose sight of overall programme shape and content, they will often call their own teams together at various times as their plans begin to crystallize. A preliminary canter through items and their treatment is always useful, but often the pivotal part of any newsroom schedule is a meeting to structure in a more formalized way the sequence in which items are to be transmitted.

Without this information readily to hand in some consistently acceptable form, the production staff would be unable to translate editorial wishes into a televisual format, and the result would be obvious to the viewer in a short time. Presenters would be addressing the 'wrong' cameras, introductions would not match the reports for which they were constructed, graphics would be out of sequence or fail to appear on the screen at all. In short, the whole broadcast would disappear in confusion.

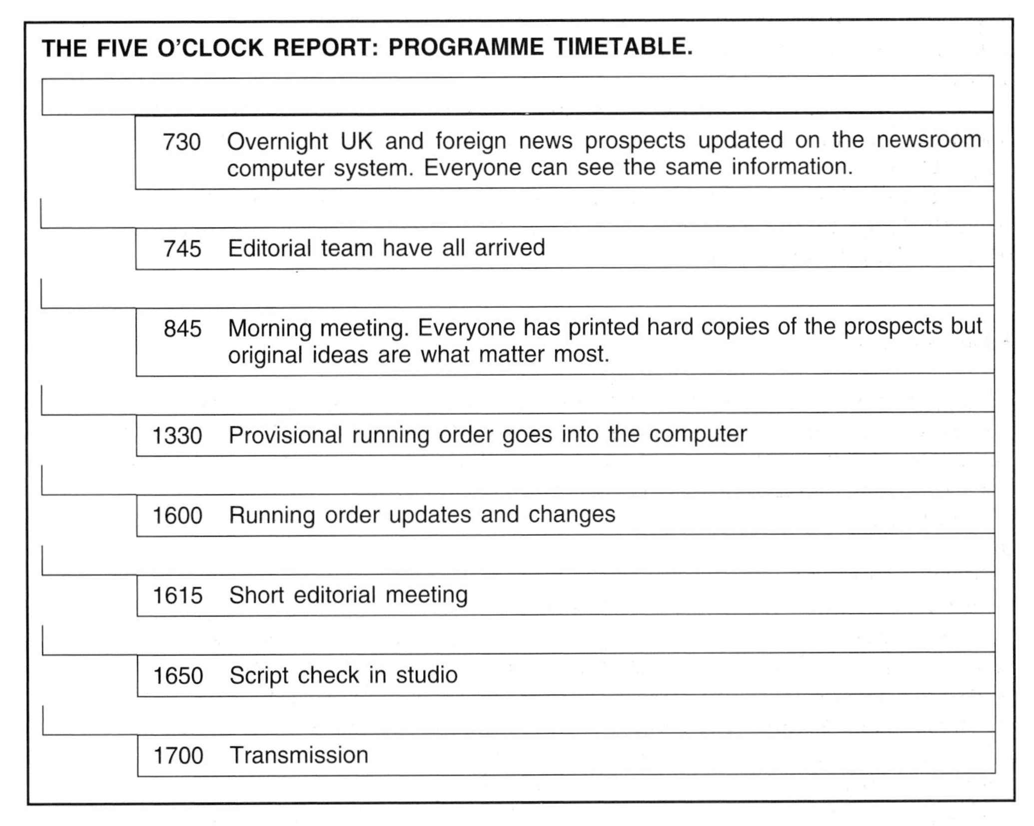

Figure 12.1 How a day on *The Five O'Clock Report* might progress towards transmission. The morning editorial meeting would involve staff on all programmes. Those reporters and camera crews not already on assignment would be briefed as early as possible: coverage intended only for *The Five O'Clock Report* might not be decided until the programme meeting at 0930. Compilation and writing of programme items begins immediately the programme prospects have been prepared, and may not finish until transmission has begun.

Yet crucial though it is to the smooth transmission of a programme, there is nothing especially complicated about this running order or rundown. In some small news operations it need be no more than half a sheet of paper on which the director notes the details as dictated by the programme editor, and then passes them on for the relevant staff to carry out. In other cases it is a lengthy, detailed document which emerges only after earnest discussion.

In either case, while practical necessity demands that it may have to be drawn up and distributed in plenty of time for transmission (and it follows that longer programmes need longer running orders, which therefore take proportionately longer to compose) the chances are it will be subjected to considerable change as programme deadline nears. That is inevitable, because even without the need to accommodate newly breaking stories, so much of what is committed firmly to the news processing computer at any stage up to and probably including transmission is no better than educated guesswork, based as it is on frequently sketchy information about such unpredictable matters as the estimated arrival of satellite pictures and sound from some remote spot, the duration and substance of a political speech (combined with the ability of the team on the spot to identify the most important part of it and feed it to base), the willingness of an interviewee to postpone a date in a restaurant in favour of two minutes in front of a robotic camera in an otherwise unattended television studio, added to the notional time it takes to edit and script half a dozen complicated packages.

For all this, the running order is the foundation on which any programme begins to assume a definite shape, and even experienced editors feel uneasy until it has been prepared. Many writers, meanwhile, like to know how the elements for which they are responsible are meant to fit into the overall scheme of things, if only because it enables them to decide whether they will need to construct phrases linking one item to the next.

The programme editor/producer will probably preside over the running order conference, perhaps canvassing colleagues for their opinions when outlining the framework and giving an idea of the approximate time he or she proposes to devote to each separate item. In some longer news programmes, departmental heads with their own teams representing consumer affairs, environment, sport, arts etc., contribute regular blocks, having a fixed allotment of airtime without needing to bargain for space.

This semi-autonomy of specialist areas can create enormous difficulties for editors trying to balance editorial priorities as they see them, and often represents a weakness either in the system or the people involved. Another headache for senior programme heads is the interest sometimes shown by executives even higher up the scale, especially those who find it difficult to keep their hands off operational areas they once controlled. This happens in all news areas. It is not unknown for a daily routine, originally established as a genuine wish by a senior executive to be informed about how things are going at the sharp-end, to metamorphose with time into yet another full-blown editorial conference which keeps editors away from their main duties for too long and during which 'suggestions' made about the running order or individual items cannot be ignored. Other editors will complain that distance is no saviour. Thanks to modern communications systems, senior executives need not leave the comfort of their own homes or offices to view programme running orders and if necessary take issue by e-mail or telephone with those notionally in charge.

Along with the morning editorial meeting, the running order conference, held at an unvarying point in each day, should represent one of the rocks on which the programme infrastructure is built and be mandatory for all staff involved to attend unless engaged on matters so essential that the team effort would otherwise be affected negatively.

RUNNING ORDER. FIVE O'CLOCK REPORT.

no.	title	source	writer	duration	cum.dur.	checked
001	TITLES/HEADS	VTS	JJ	0.45		
002						
003						
004						
005	ELECTION/INTRO	CAM 1/INSET/STILL		0.45	01.30	
006	ELECTION/SCENES	CVT/SUPERS	CR	1.00	02.30	
007	ELECTION/REACTIONS	CAM 1/INSET		0.30	03.00	
008	ELECTION/INTVWS	VT/SUPERS		1.30	04.30	
009						
010						
011						
012	ELECTION/REST	CAM 2/CU/CHARTS		0.45	05.15	
013	ELECTION/ROUNDUP	CAM 2/CU		0.15	05.30	
014	ELECTION/POLITICAL	OB/SUPERS		2.15	07.45	
015	ECONOMY	CAM 1/INSET	JL	0.15	08.00	
016	ECONOMY/MIKE	CAM 3/CHART				
017	ECONOMY/WRAP	VT/SUPERS		3.00	11.00	
018						
019						
020						
021						
022	DAMAGES	CAM 2/INSET/MAP	RS	0.45	11.45	
023	DAMAGES/INTERVIEW	VT/SUPERS		1.00	12.45	
024	DAMAGES/2-WAY	OB/SUPERS		2.15	14.30	
025	DAMAGES/BACKGRD	VT/SUPERS		1.30	16.00	
026	DAMAGES/TRAIL	CAM 1/INSET	TC	0.15	16.15	
027						
028						
029						
030	HALFWAY	CAM 1/CU	TC		16.30	
031	FARMING	VT			16.45	
032	ORPHANS	VT			17.00	
033	RECORD	VT	RB		17.15	
034	CRIME FIGS	CAM 2/INSET	MR	1.00	18.00	
035	CRIME/NW	VT/SUPERS		1.45	19.45	
036	CRIME/MET	CAM 2/INSET/CHARTS		0.30	20.15	
037	CRIME/CHIEF	STUDIO 1 + 1/SUPER	JS	2.15	22.30	
038	KIDS	CAM 1/CU	RA			
039	KIDS/SMITH	VT/CHARTS/SUPERS		2.30	25.00	
040						
041						
042						
043	TOPFIELD	CAM 2/CU	MR	0.15		
044	TOPFIELD/EUROPE	CVT		0.30	25.45	
045	TOPFIELD/REPORT	VT/SUPERS		3.00	28.45	
046						
047						
048	CRICKET	CAM 1/INSET	JG	0.15	29.00	
049	CRICKET/AUSSIE	VT		1.00	30.00	
050						
051						
052	END HEADS	CAM 1/CU				
053	END/PIX	VTS/SUPERS		0.45	30.45	
054	CLOSE	CAM 3/END TITLES		0.30	31.15	

Figure 12.2 The running order in the computer compiled after *The Five O'Clock Report* programme meeting. Different news organizations lay out running orders in different ways but this is typical for any of them. Every item is allocated one or more page numbers and titles to ensure flexibility. Other columns may show technical resources and the name of the writer/producer/reporter or an embargo time. The notional allocation of time at this early stage gives the editor an indication of overall running time. Changes to content and duration are inevitable.

The rest of any typical programme timetable (see Figure 12.1) is very much a matter of taste, dependent as it must be on factors which vary from organization to organization. Logic suggests, however, that a brief review of the running order is useful before the production team repairs to the studio, and that pre-transmission checks (see pp. 190–191) are equally valuable.

Running order format

Some programme heads leave their production staffs with the responsibility for constructing and amending running orders, but as so much depends on editorial as well as technical judgements there remains a general reluctance on the part of most journalists to relinquish this control.

The idea of establishing a formal running order format would be considered almost laughable by many newsrooms for whom the daily newscast is a simple, almost mechanical matter. For others, the benefit of consistency is obvious. Bigger newsrooms with regular staff turnover need newcomers to understand and comply unfailingly with a system which leaves no room for error.

Insistence that running orders and scripts are constructed by everyone to a carefully defined format makes for speed and efficiency. With the introduction of uniformity so many simple mistakes are avoided: in the understandable haste which accompanies any approaching deadline it is all too easy to go wrong when no one is entirely sure whether the item entitled 'Politics' in the running order is the same as the one headed 'Downing Street' on the script or whether it is the same as the 'Tax Interview' tape which has been delivered to the transmission suite. It takes no great leap of imagination to appreciate how easily the wrong item could be transmitted.

Fortunately, with modern technology, mistakes are less prevalent than they once were. One of the main strengths of any of the newsroom computer systems is the ability to cope easily with the creation of running orders, and to handle all possible combinations of changes, up to and including transmission, within seconds.

We will come to that later. For the moment let us concentrate on the example of a fairly typical running order (Figure 12.3). Each item is given a page number and title by which it is identified during its brief life as a way of avoiding the kind of confusion described above. Where stories are particularly complex, or have several strands to them, they are allocated successive page numbers and titles. This is so that the individual writers responsible for them can produce scripts page-by-page to speed distribution.

Our specimen running order also contains several numbers against which there are no titles. This is a built-in allowance for any new stories to be slotted in without disturbing everything else. Where really drastic changes cannot be accommodated the choice may be between scrapping whole pages and renumbering or accepting pages out of numerical sequence. Whatever the decision employed, any changes, however minor, have to be communicated to the production and technical staff, so by the time the news begins no one is in any doubt about the part he or she is meant to play.

While the running order layout itself is very important, it does not have to be complicated. The chief consideration is for it to be easily understood by all. Essential ingredients are numbers, titles and sources, and behind what may seem to be a straightforward exercise, a great depth of journalistic experience and understanding has to be shown by the editor in the hours between the round of formal and informal conferences and the drawing up of this instrument.

RUNNING ORDER. FIVE O'CLOCK REPORT.

no.	title	source	writer	duration	cum.dur.	checked
001	TITLES/HEADS	VTS	JB	0.45		
002						
003						
004						
005	ELECTION/INTRO	CAM 1/INSET/STILL		0.45	01.30	ap
006	ELECTION/SCENES	VT/SUPERS	CR	1.03	02.33	ap
007	ELECTION/REACTIONS	CAM 1/INSET		0.32	03.05	a
008	ELECTION/INTVWS	VT/SUPERS		1.30	04.35	
009						
010						
011						
012	ELECTION/REST	CAM 2/CU/CHARTS		0.48	05.13	
013	ELECTION/ROUNDUP	CAM 2/CU		0.18	05.31	a
014	ELECTION/POLITICAL	OB/SUPERS		2.15	07.46	
015	ECONOMY	CAM 1/INSET	JL	0.15	08.00	
016	ECONOMY/JONES	CAM 3/CHART				
017	ECONOMY/WRAP	VT/SUPERS		2.53	10.54	ap
018						
019						
020						
021						
022	DAMAGES	CAM 2/INSET/MAP	RS	0.40	11.34	ap
023	DAMAGES/INTERVIEW	VT/SUPERS		1.00	12.34	
024	DAMAGES/2-WAY	OB/SUPERS		1.45	14.19	
025	DAMAGES/BACKGRD	VT/SUPERS		1.30	15.49	
026	DAMAGES/TRAIL	CAM 1/INSET	TC	0.15	16.04	ap
027						
028						
029						
030	HALFWAY	CAM 1/CU	TC		16.19	
031	FARMING	VT			16.34	
032	ORPHANS	VT			16.49	
033	RECORD	VT	RB		17.04	
034	CRIME FIGS	CAM 2/INSET	MR	0.50	17.54	a
035	CRIME/NW	VT/SUPERS		1.45	19.39	
036	CRIME/MET	CAM 2/INSET/CHARTS		0.30	20.09	ap
037	CRIME/CHIEF	STUDIO 1 + 1/SUPER	JS	2.15	22.24	
038	KIDS	CAM 1/CU	RA			
039	KIDS/SMITH	VT/CHARTS/SUPERS		2.21	24.45	ap
040	JOYRIDERS	CAM 2/MAP	FS	0.15	25.00	
041	JOY/MOTORWAY	VT/SUPERS		0.45	25.45	
042						
043	TOPFIELD	CAM 2/CU	MR	0.15		
044	TOPFIELD/EUROPE	VT		0.30	26.30	
045	TOPFIELD/REPORT	VT/SUPERS		3.02	29.32	ap
046						
047	PICASSO	CAM 1/INSET		0.15		
048	CRICKET	CAM 2/INSET	JG	0.15	30.02	a
049	CRICKET/AUSSIE	VT		1.00	31.02	
050						
051						
052	END HEADS	CAM 1/CU				
053	END/PIA	VTS/SUPERS		0.45	31.47	
054	CLOSE	CAM 3/END TITLES		0.30	32.17	ap

Figure 12.3 The running order after the 'Joyriders' and 'Picasso' stories have been added. The 'duration' and 'cumulative duration' columns now include adjustments to take account of known changes in item length. Although there have been some minor trims – including the reduction of page 24 by 30 seconds – the programme is still theoretically more than three minutes over its scheduled 29 minutes 30 seconds, and decisions about the substantial deletions necessary have yet to be made. 'a' in the final column indicates that the script has been written and approved; 'p' denotes that it has been printed and distributed. By the time the programme is broadcast the rest of the running order will be marked in this way. Some programmes have 'approval' boxes (read by the editor and cleared for transmission) with a simple tick mark.

Timing

One of the most important factors in the construction of any running order in a fixed duration news programme is the allocation of a notional length to every story and, where multi-element stories are concerned, a notional length for each component. It is both an intellectual and journalistic exercise. All allocations are based on what weight an editor attaches to the interest and importance of separate items, their treatment, provisional position in the order, and the effect each element contributes towards the content and tempo of the programme as a whole.

Time is not the only consideration. Short bulletins are just as difficult to compile and require the same attention to detail, even though they may lack some of the trimmings associated with lengthier productions. No part of the programme, however seemingly insignificant, must be missed. Every facet has to be built in and accounted for as part of the tally, including opening and closing title sequences, headlines if any, and the time it takes for the presenters to ad-lib their sign off.

At this stage the arithmetic need only be very rough, but without it the editor will have no idea whether his or her programme is likely to meet its target duration. This, within the boundaries of conventional programme scheduling, is almost certain to be consistently very strict to meet overall station requirements. These may encompass 'junctions' with complementary channels, advertising breaks and the transmission of regional or local programmes. So any production of any type which does not carry advertising will probably be a minimum of thirty seconds shorter than its supposed length, the spare time being used for in-house 'trails', and other presentation announcements as a means of ensuring the next programme, of whatever category, starts exactly when scheduled. In this way, then, a typical half-hour news programme may last only 29 minutes 30 seconds, while a commercial half-hour news is likely to be nearer 22 minutes by the time advertisements are taken into account.

Another consideration to be faced by the editor in compiling a running order is the relentless pressure generated by the uncertainties of news and the process of gathering it. So much can go wrong, with stories going down for editorial or technical reasons, or simply not turning out as planned, the nightmare prospect of under-running is ever-present. It takes a steely nerve born of vast experience (including a few near-disasters) not to over-commission for the sake of safety or to succumb to the temptation of being too generous with material which comes in early and is available.

Editors known for a consistent tendency to over-set earn a justifiable reputation for being indecisive, and the inevitable consequence – having to leave out stories from the final transmission list – often leads to disappointment and anger on the part of those who might have sweated blood to gather material which ends up being junked for no other reason than it is fortuitously the 'right' length to fit the editor's eventual need for savings in timing.

Some editors, mindful that a story dropped amounts to wasted effort and resources, build in ostensibly timeless 'shelf' or 'standby' edited items as an insurance against some sudden news famine, in the confident expectation it will never happen. In theory it seems a good idea. Unfortunately it has been known for weeks to go by with the same report dutifully included in every running order and put back on the shelf unused, until the day comes when – against all the odds – it has to be dusted down and drafted into emergency use. Because it has been previously viewed and found acceptable, no one in the hard-pressed news team feels the need to do so again, and on transmission the 'shelf' item is found to be embarrassingly out of date or to include an interview with someone who has since changed their opinion or their job or simply died.

There are other influences at work, too. In the competitive world of television journalism, reporters and correspondents are not above pushing their own stories against those of others, not hesitating, if necessary, to go over the head of the editor to more senior figures in the hierarchy as a means of encouraging favourable treatment. Insistence from a specialist correspondent that this or that report 'must go today' because the opposition may have caught wind of it is hard to resist, as are the exhortations from the foreign desk that a news team's expensive deployment somewhere or other abroad is sufficient reason for every contribution they send to be used, never mind how little news they have to report.

Editors certainly do have to take all this into account, but they also have to be prepared to battle for their programme as they visualize it and on behalf of the rest of the team helping to shape it. Because it happens to be a star reporter who is pleading for 'just an extra thirty seconds' on a piece there is no reason to agree automatically, especially when 'just an extra thirty seconds' are being sought by a dozen others, perhaps with more justification. Editors have – should insist on having – the last word, and those who give in too readily are soon marked down as an easy touch by the bullies and lose the respect of those whose stories have to be sacrificed. A disciplined approach to the process must at the same time not rule out a willingness to accept changes in content, treatment and duration of any item, where these seem appropriate. Flexibility and confidence based on secure judgement about news values in general, as well as the worth of every item to the programme, should be a prerequisite for appointment to the role of an editor: and if that means putting older or more experienced journalists in charge, so be it.

Being entirely responsible for what perhaps several million people will see on their television screens for a short period is a thrilling and privileged proposition for those who care to think about it, so the editor will attempt to bear all this in mind when roughing out a first draft.

For news editors, making notional timings in multiples of 15 seconds will make the arithmetic simple at this stage: adjustments can be made when the real figures are known. Not even the most optimistic editor expects the duration of every story to be as originally allocated, but the tendency is for some of those which end up slightly longer to be offset by some of those which end up slightly shorter. Later, as the hours count down towards transmission and the duration of every finished item is fed into the newsroom computer system, it is hoped the running order will show a balance being achieved, the aim being to make it unnecessary to discard large chunks of material.

Dealing with breaking news

In addition to monitoring the progress of items on the running order as first envisaged, the editor's antenna should also be tuned to the possibility of unexpectedly 'new' news.

For journalists who thrive on the kick of late-breaking news, television offers endless opportunity to indulge themselves. Scarcely an hour goes by in any newsroom without some event worthy of being reported presenting itself. The predicament facing editors is what to do about it. And when.

The television news machine, for all its ability and sophistication, is in many ways still very cumbersome. Putting a newscast of any size (duration) on the air is not unlike putting something into space: you can stop it being launched right up to the last moment, but once the thing is on its way it is very difficult to change course significantly without jeopardizing the entire mission.

On a network newscast the number of editorial, production and technical staff who need to be informed of every minute change to the running order will probably run into scores, and while most straightforward cuts and additions are accommodated routinely with practised ease, the warning bells may start to ring once alterations call for, say, the unexpected promotion of a middle-order story to the top, with all the attendant knock-on effects. The closer the changes have to be made to the time of transmission the greater the scope for misunderstanding, with confusion and uncertainty almost certainly leading to disaster on screen. Computers are all every well, but the determining factor is the ability of humans to communicate successfully under what can be extremely stressful conditions: psychologists would have a field day studying the behavioural patterns of those working in a newsroom or programme control room when big and late changes to a running order are taking place.

No sensible editor who has been this way before will ever choose to make amendments later than necessary, and will do so only after careful thought, mindful of what can go wrong. In some, the prospect of the whole thing falling apart on screen engenders a timid approach which rules out all but the most minor alterations beyond a self-imposed deadline – perhaps an hour before transmission – their argument being that a smooth, trouble-free production is preferable to one which risks being noticeably ragged for the sake of editorial changes of questionable worth. This view is inclined to change only when circumstances suggest it would be journalistic suicide not to do so.

Some programme running orders are fashioned with such care and thought, with one item flowing logically into the next, that editors are reluctant to accept changes or additions which would appear not to fit. At the other end of the scale are the gung-ho merchants who seem incapable of making up their minds and are forever tinkering with the programme right up to and usually including transmission, raising blood pressures (their own included) all round.

Countdown to transmission: *The Five O'Clock Report*

As a way of gaining some idea of how modern television news reaches the screen, we'll use the basis of experience gained from several real programmes and apply it to the crucial last few hours before transmission of an imaginary example. We'll call ours *The Five O'Clock Report*, a daily half-hour, advertisement-free national news programme aimed at a tea-time audience. Its style is serious and significant – bordering on the broadsheet – and its brief is to concentrate on the main stories of the day. While foreign coverage is certainly not ruled out, the emphasis is deliberately towards the domestic scene. News programmes later in the evening will be expected to broaden the agenda.

1.30 p.m.

By this time of the day, the editor's thoughts about the content of the programme he will lead to transmission three and a half hours from now are fairly clear, based as they are on the knowledge of what his colleagues and the opposition have already put out in their breakfast and lunchtime news. The compilation of his own programme is based on three factors – news which has not changed but is still worth carrying over from earlier; developments giving a fresh look to other stories which have also been previously carried, and the few new assignments intake has planned to cover during the afternoon. The fourth factor, breaking news, is something over which he has no control.

The journalists allocated to him have spent the morning working on some of the elements he had already expected to include, but when they come together to put the components of the programme on paper, members of the team are unable to add anything significant to the informal discussions they have been having sporadically over the past few hours. Nothing that has happened encourages the editor to rethink the programme plan he started to form not long after reaching his desk first thing.

So, in the absence of anything more startling, a political story leads the way. Two overnight by-election results have raised questions about the leadership of the governing party. It is undoubtedly an important story, but unfortunately some of the shine has been taken off by the predictable nature of the outcome, and its place at the head of all bulletins since breakfast. Instinct tells him that if it is to continue to stand up as the lead item, it must be treated as substantial and multi-faceted. Immediately he is struck by the potential timing problems.

First, it will be necessary for the programme presenter to introduce the item by giving the main details in the studio. That is likely to take 45 seconds. A tightly edited selection of the noisy scenes when the results were announced will take about a minute, and interviews with senior representatives from the winners and losers are unlikely to make sense in anything less than a total of a minute and a half. Round off the story with other reactions and an assessment from the political editor, add the obligatory 45 seconds for the opening title sequence, and 7 minutes 45 seconds will have already gone by.

Second place in the running order goes to the economy. Oddly, in direct contrast to the government's electoral failure on the previous day, the latest inflation and other indicators show signs of a continuing upturn in the country's fortunes. In the editor's mind it would be perverse to ignore the logic of following the one with the other. However, although the figures in themselves are easy to explain, the economics correspondent insists that his specialist gloss, illuminating the trend with some fancy graphics, can be applied only at the cost of two and a half minutes of airtime.

With this, the arithmetic indicates more than eleven minutes will have been accounted for.

At one stage the editor was tempted to lead his programme with a more popular human interest story which is sure to make all the next day's newspapers. The survivor of a serious road accident 18 months earlier has been awarded record damages against the driver of the car which injured her, and the prospect of an interview with an articulate young woman making remarkable progress towards resuming a normal life against all medical expectations would have had particular appeal. The story has the extra attraction of being new since lunch, but the disappointing word from outside the court where the case was heard is that *The Five O'Clock Report* camera team were jostled during the media mêlée on the pavement and although what the interviewee said can be heard clearly the reporter believes the pictures are barely steady enough to sustain half a minute.

So it is with some reluctance the editor decides to put the story third: although the interview is disappointing he is promised an interesting minute and a half 'two-way' with the reporter who covered the case and a short backgrounder on the surgical techniques employed to put the injured woman on the path to recovery. This particular item has been extracted from a half-hour documentary made by another section within the news department, and the editor makes sure to allow an additional 15 seconds to 'trail' the programme, which is due for transmission later the same evening.

If all goes according to plan 17 minutes 30 seconds will have elapsed by the end of this story.

At this point, about halfway through the programme, the usual style is to run a brief sequence of pictures reminding viewers of the main stories still to come. This device serves much the same purpose as the commercial break in other news programmes: it helps separate the harder-edged 'top half' from the slightly softer items later on.

The 'midway heads', although useful, can also prove to be something of a headache. First, they use up nearly 45 seconds of precious airtime. Second – and more important – they reduce the editor's flexibility to omit material if he runs into timing problems later on. Little looks more foolish than trailing a piece which is eventually forced out of the newscast, so he chooses his three items very carefully. The sequence will begin with a shot of the cows from the Topfield Farm report, an inner-city street scene to illustrate an item which appears to establish a definite link between youth unemployment and crime, and – on an entirely different plane – a boundary being struck by a relatively unknown Australian cricketer who has equalled the record individual score in a test match. By some means these extracts – plus the stories they represent – will be retained, whatever else has to go in an emergency.

The biggest uncertainty of the day is over another, related item in this part of the programme. Somehow it had slipped intake's notice with the result that the planners have only just become aware that a new report containing the latest official crime statistics is due for publication later in the afternoon. It is being hinted that they will show an overall rise in clear-up rates by the police, but although copies of the report are being issued to specialist correspondents within the next hour, for reasons the editor cannot understand it is embargoed for use until long after his programme is over. There has been, he believes, some kind of bureaucratic mix-up which he is urging his planners to sort out before the programme.

In the meantime he wants a substantial item prepared in the hope that the embargo will be lifted, and he also wants an interview with the author of the report. If necessary he is prepared to do it live in the studio. This whole segment, he believes, could move up the running order, but for the moment, on the premise that figures do not show a dramatic improvement – or officialdom would be shouting about them – he is content to leave it where it is in the running order.

The problem is if the crime story does make it – by the time he has added the remaining stories and the closing sequence he will, in theory at least, already be exceeding his allocated half-hour (29 minutes 30 seconds) by nearly two minutes.

As they go through the items in the planned order of transmission, the studio director makes notes and asks questions about the likely production requirements. The opening and closing sequences are standard, the use of three studio cameras, the need to create insets behind the presenters on some stories and the live link to the political editor are all routine. The only slight difference about today is the possible inclusion of a studio interview, which with this programme usually happens no more than once or twice a week. The director is keen to know at what point in the half-hour the author of the report is expected to appear – it will be necessary to get the interviewee physically into her seat at an appropriate moment. The camera movements will have to be checked.

2.15 p.m.

With a running order in front of him on his screen, and on hard copy in his hand, the editor feels more at ease. The important decisions have been taken, everyone knows what is expected of them, and now there is a little time to relax and reassess.

Although theoretically he has too much material, he considers it to be easily manageable at this time of the day. He has already decided that if the crime figures and the interview do become available he will probably discard the medical element of the 'damages' story. The trail would still be relevant, but dropping the background minute and a half would make a useful contribution toward the two minutes he needs to save, and the rest could be made up by judicious editing of other elements of the programme.

Nevertheless, looking again at the running order, he considers the programme overall to be too 'heavy', lacking in the shorter items which would improve its pace. There is still time to do something about it.

2.35 p.m.

The programme has two regular presenters, a man and a woman, who 'lead' alternately. Both would normally have been at the afternoon meeting, and at least one should have been working alongside the editor in the newsroom since mid-morning, offering suggestions and becoming familiar with the items to be read later on. Today, though, the female presenter has been unavoidably absent, reading the hourly news summaries in place of a sick colleague. The male presenter, an experienced former correspondent, has been involved in a station publicity event for a forthcoming new series of factual programmes he has been asked to introduce, and has been delayed by an interview with an inquisitive feature writer from the regional press.

Now he arrives, full of apologies, needing to catch up. He has not seen or heard any news since early morning and feels uncomfortably out of touch with events. He and the editor go quickly but thoroughly through the running order, discussing each item in turn. Extra attention is given to two elements: the content and duration of the 'two-way' to be conducted with the reporter at the scene of the 'damages' story, and the possible studio interview with the author of the crime figures report. Agreement is reached about the general form the questions will take, and then the presenter leaves the newsroom. In the next three-quarters of an hour he needs to take a quick look at a recording of the lunchtime news and research the background of the author of the crime report.

2.55 p.m.

The foreign assignments editor comes over to the newsdesk. Because of earlier technical problems, a 60-second sequence of pictures showing the record-equalling test match score, made against India, has only just become available. They are now scheduled to come in by satellite from Sydney in two minutes' time. The editor takes a personal interest in cricket, and advanced newsroom technology will allow him to enjoy the pictures without leaving his desk. He picks up a pair of headphones and switches channels on the small television monitor by his elbow to the incoming satellite signals. He knows the link with Sydney is already established because he can hear 'tone' (an unbroken signal used to identify sound sources) and see colour bars.

A voice breaks in on the desk intercom. The journalist working on the Delia Ward farming package has nearly finished putting it together: would the editor like to come to the editing suite to see it? 'I'll be along in five minutes', says the editor.

Tone on the satellite signal has now been replaced by a repetitive message. 'This is the international sound and vision circuit from the Channel 86 Broadcasting Corporation in Sydney, Australia. This is the international sound and vision circuit from the Channel 86 Broadcasting Corporation in Sydney, Australia. This is the international sound and vision circuit from the Channel 86 Broadcasting Corporation in Sydney, Australia. Transmission will begin in 30 seconds. Please start your recording machines. Transmission will begin in 30 seconds. Please start your recording machines. Transmission will begin in 30 seconds. Please start your recording machines.'

A clock with a moving second hand takes the place of the colour bars. After 30 seconds the clock disappears and pictures of cricketing action begin. The batsman is seen vigorously hitting the ball to the boundary. The next shot is of a shirt-sleeved crowd applauding enthusiastically. Unfortunately the editor cannot hear them. He fiddles with the earphones and the monitor switch. Bat hits ball soundlessly towards the pavilion, followed by more silent clapping.

The foreign assignments editor calls on the intercom. 'I'm not getting any sound on the cricket. Are you?' No, says the editor. On the monitor the batsman is still laying about him. Much more than the expected one minute of action has taken place – but still no sound. Abruptly, in the middle of the bowler's run-up, the pictures vanish. Colour bars and the repetitive message return. 'This is the international sound and vision circuit from the Channel 86 Broadcasting Corporation in Sydney, Australia. This is the international sound and vision circuit from the Channel 86 Broadcasting Corporation in Sydney, Australia. This is the international sound and vision circuit from the Channel 86 Broadcasting Corporation in Sydney, Australia.' 'They've obviously got problems', says the foreign assignments editor. 'I'll have to check with international control.'

3.10 p.m.

The editor is viewing the Topfield Farm offering in one of the five edit suites situated in the corridor next to the newsroom. His employers expect him to carry full responsibility for the programme he edits and he takes it very seriously. As a matter of principle he expects to read every script and see every item before it is transmitted. Those posing potential legal or ethical problems he will scrutinise in minute detail. Admittedly some stories will not be ready until very late or perhaps not until after the programme has begun, and there are obvious difficulties with the live contributions, but he has taken care to entrust these to the most reliable members of his team. He will in any case keep an eye on progress as late as possible. Other stories will require adjustment – sometimes drastic recasting – or may not meet his editorial expectations at all and have to be omitted. There is also the consistently thorny problem of duration. He has no intention of leaving to others a decision to let a story over- or under-run its allocation. And he cannot make that judgement unless he has the opportunity to assess how valid is that call for 'just another 30 seconds'.

He views Delia Ward's Topfield Farm story once all the way through, without comment, making notes as he goes along. She has done well, and he likes the way the pictures have been put together, but his first impression is that the interview is slightly over-long and he thinks it would be possible to save about 20 seconds. He asks the picture editor to show it again but, after some hesitation on the second viewing, comes to the conclusion that cutting it would spoil the sense of the interview. Leave it unchanged, he tells the picture editor. With that he marks the finished duration – 3 minutes and 2 seconds – in the margin of his running order, and heads back towards the newsroom.

3.25 p.m.

The studio director and the producer responsible for the 'crime figures' segment of the programme are waiting for him. They disagree about a backing for the interview, if it takes place. The original production plan is for an electronically keyed inset symbolizing the report to appear over the presenter's left shoulder when he reads the studio introduction. The producer would like it to continue during the interview: the studio director thinks it would look out of place and add to his problems with camera angles. A decision is needed. The editor gives the matter some thought. Friction between these two talented but occasionally prickly people is not unknown and he cannot risk the loser of the decision going off to sulk. The compromise solution is easy, he tells them. Drop the inset altogether.

3.30 p.m.

Signs of life from Sydney. The satellite pictures are running again. There are healthy cricketing sound effects for the first ten seconds, a loud 'clunk', then silence. Tone and colour bars return. They will have to try again, but not until four o'clock, says the foreign desk, because priority over bookings has gone to someone else. It's going to be tight.

The presenter, comfortably attired in tee-shirt, scruffy jeans and old trainers, is pecking away at his keyboard with both index fingers. He likes to shape the opening story to suit his style, and is busily turning some agency copy and notes from the item producer into viewer-friendly language. If there is time he will also write the headlines.

Intake have good news and bad about the crime figures story. The correspondent covering it has convinced the press officer for the government department responsible for the report of the benefits of bringing the end of the embargo forward so that the item can be included in the programme. But the author of the report is not available for interview. The correspondent offers to write a piece and appear in the studio instead. The editor is not enthusiastic: he has already scheduled live contributions from two specialists – one from the political editor, the other from the economics correspondent. Go back and find out why the author is not available, he orders intake. Surely she must be aware of the importance of the report.

3.50 p.m.

Another viewing. This time the report of the 'damages' story is coming in over the microwave link from near the court. The interview with the young woman is really very good. She is composed, dignified and forgiving towards the driver who injured her, and somehow the unsteady pictures of the unedifying media scramble add a moving counterpoint. The editor likes what he sees. He is tempted to move the story up the running order after all. He gets on the intercom to the producer in the editing suite where the pictures are being recorded. The story is in danger of being underplayed. If these are the raw rushes from the camera and not an edited version of the scenes after the case he would like more of them.

Other stories are being put forward for consideration. Reuters reports that thieves posing as picture restorers have made off with a priceless Picasso from a Dutch museum. A journalist from one of the regional newsrooms telephones to say two teenaged joyriders have been severely injured on a motorway, crashing the car they stole from a service station. Some helicopter shots of the resulting traffic chaos may be available by five o'clock. Coastguards warn that a freighter carrying a consignment of sheet metal is in danger of going aground in heavy seas off Shetland.

Yes – as long as we can get a picture of the picture, says the editor. He knows he will annoy his viewers if he runs a story about a stolen painting without having a picture of the painting! Definitely yes to the joyriders. Keep an eye on what happens to the freighter.

4.05 p.m.

The author of the crime figures report has been located a hundred miles from the studio. She is at home packing for a trip in connection with her report, but will reluctantly agree to interrupt her preparations to do an interview. There is not enough time in which to dispatch a reporter from base, so the plan is to ask the nearest regional station to provide one. The snag is it is not yet clear whether a camera will be available as well.

Sydney is up again, for the third time of asking. The editor puts on his headphones. With relief, he sees the pictures and hears the sound of bat on ball. But there is something else wrong: he was promised the voice of an Australian commentator. Commentary is like part of the action. Where is it?

The correspondent covering the crime figures story wants to know – slightly irritably – whether or not his services are needed. The editor is in a dilemma. He hesitates. Give me ten minutes, he says.

The studio director bustles into the newsroom on his way to the studio control room. Anything new? The editor leans back in his chair. Changes are imminent but as usual they depend on matters beyond his control. Decisions will have to be made very soon.

4.15 p.m.

Three-quarters of an hour to go, and still the programme is not finalized.

The regional station has confirmed it does not have a camera within fifty miles of the author's home, but intake do have an alternative suggestion. She can be driven by taxi to the local studio and interviewed live, 'down the line'. The intake editor doesn't feel it necessary to add that the journey takes at least half an hour even on a good day, but, fingers crossed, the interviewee should still be there well in time for five o'clock. The editor says yes, and the correspondent is told his services are not required.

Helicopter pictures taken after the joyriders' crash show up on the monitor. The motorway is impressively choked. It's looking a better story than was first indicated. The editor toys with the idea of moving it up the running order.

At last Val, the female presenter, looks into the newsroom to say she has finished her long stint reading news summaries all day, and starts to get up to date with the scripts. At least she is well 'up to speed' with most of what has been happening over the previous few hours.

For the past ten minutes her co-presenter has been in one of the edit suites, looking at pictures which will make up the headline sequence. Now he returns to scoop up more of the completed scripts from his in-tray before going off to join the director in the studio for whatever rehearsal will be possible. His tee-shirt has now given way to a smart jacket, shirt and tie, but the editor notes with amusement that the scruffy jeans and old trainers are still there.

Setting out the script

Hand-in-hand with the numbered running order goes a need for the careful layout of every page of script. This discipline is perhaps not vital in offices where the same very small editorial and production teams are at work every day, but big services with lots of output, especially those operating along shift-working lines, need to establish a consistent house style readily understood by everyone. The whole purpose is to ensure that every newscast proceeds smoothly, without the embarrassing on-screen glitches which detract from the production.

Just as the running order dictates what elements the programme contains, the written script includes indicators as to how and when they are meant to be introduced. The studio director, glancing for the first time at a page arriving in the control room two minutes before airtime, must be able to feel absolutely confident that following the instructions without hesitation will not lead to disaster. Poor typing or idiosyncratic layout could easily lead to misunderstandings among the production staff, with predictably disastrous consequences for live programming.

Scripts should be produced by individual members of the editorial team according to the convention which says each page must be typed onto one page, in effect, one page on a computer's script segment. Technical instructions are in one colour or font (usually red or italic or both) and the script is in black. The technical instructions dictate what the viewer sees. The script in black is what the viewer hears.

The number of script copies required for each newsroom depends on the number of people most concerned with the production. At the very least the presenter(s), technical director, programme editor and those responsible for sound and graphics should all see the same thing on their screen and be able to read their own areas of

Fives O'Clock Report	**Titles/Heads**	**Mon Jun 13**

[Live Read + Sig Tune]

VT OOV (Elections)
Duration 8"

Defeat for the government in two crucial by-elections brings new talk of a leadership crisis.

VT OOV (Damages)
Duration 7"

The woman maimed by a motorist says she forgives him.

VT OOV (Crash)
Duration 7"

A motorway choked after joyriders crash the car they stole.

VT OOV (Cricket)
Duration 8"

And a virtual unknown bats his way into the record books.

RUN SIG STING

Figure 12.4 The script as it might look for the opening of *The Five O'Clock Report*, with production instructions and text in the journalist's computer screen. In electronic news systems layout and styles will vary between different organizations, but they all follow this basic format. 'Live Read' means just what it says – the presenter is live and talking as the pictures for the headlines are shown on screen. 'VT OOV' means *Videotape is playing and the presenter is Out Of Vision*. Each headline, representing the stories the editor considers to be the most important or interesting, is identified separately. Note the commentary over each is shorter than the duration available. This ensures that words for one headline do not spill over into the next, and there is time at the end of the final one for the director to move smoothly to the following item on the running order before the pictures run out.

Five O'Clock Report	Elections Intro + CapGen	Mon Jun 13

[Live Read + Inset/Opposition Victor]
[John In Vision]

Good Afternoon. Less than 24 hours after the government's defeat in two crucial by-elections, criticism of the leadership is already being voiced. Several senior back-bench Members of Parliament have called for an immediate inquiry into the way the Party is organized. But the Prime minister has insisted there's no need to panic. Addressing a meeting of Party agents in London this afternoon he said
[Capgen + Quote]
[John oovs]
Setbacks in by-elections, however hard we try to avoid them, should come as no surprise. The important thing is we still have our majority and will continue to govern.
[Live Read]
[John In Vision]
Last night's defeats bring the government's working majority over all other parties to fifty. Opinion polls had shown the Opposition well ahead in both constituencies up to polling day. But few had forecast the scale of their success.
[Duration: 45"]
[Automation/Run VT/Jones

Figure 12.5 A generic script for the first of two pages allocated to the opening item. The script indicates that a video insert by a reporter called Jackson is to follow, but the details will appear separately, once work on it has been completed. Words in [brackets] are the technical instructions. This is what the viewer sees while the script is what the viewer hears. [Live Read + Inset] means just what it says – the presenter [John In Vision] is in vision and talking to camera, with a snapshot of the Opposition Victor at the presenter's shoulder either left or right. [Capgen + Quote] and [John oovs] means a picture of the subject of the news appears on screen and John briefly goes out of vision while a direct quotation from the person in the caption appears on screen. Then John comes back into vision again. The final part of the process is to play the videotape report from our reporter called Jones. The technical instructions in the [brackets] are selected by the writer from pre-set instructions in windows programmed into the electronic news system.

responsibility. News computer systems have changed the attitude towards paper, but there is no doubt that many people feel less than comfortable with only a screen to guide them and nothing else. Whatever the merits of the paperless office it is far better to have hard copies, and too many of them, rather than too few. If it is a paper as well as a screen programme then pages should be printed out and distributed as soon as they are completed, and full sets collated for reference afterwards.

Some news organizations responsible for several news strands each day may give each its own paper colour code, or number code. In this way there is no chance of a rogue script page left over from, say, the afternoon news, creeping into the late evening programme by mistake.

The best layouts are clear, simple, and easily understood by everyone. The amount of detail included is a matter for programme preference, although some information must be considered mandatory: the number and title of every script, corresponding exactly with those given on the running order; the duration, precise to the second, of separate components; the opening and closing words spoken on video or audio tape.

For the rest, a writer setting out an instruction intended to result in the presenter's face appearing on the screen might not be expected to have to say how the shot should be framed. The instruction when to introduce a graphic need not go beyond identifying the subject. In these cases the visual interpretations are left to the director. The job of director has changed considerably in recent years because of the ability of newsroom computer systems to act on production instructions in the script, without the need for human intervention. This is all about computer codes – usually a sequence of letters and digits – which enable computers to talk to each other (in the way the fictional but plausible R2D2 talked to other computers in *Star Wars*). So if the news computer in front of the journalist says it wants a name caption (capgen) at 12 seconds into a report, it can tell a computer that makes captions to do just that. The scope for error is enormous, especially if fewer and fewer people are available to check each other's work. The expression for this is WYSIWYG (whizzywig). It means, of course: What You See Is What You Get!

Five O'Clock Report **Elections/Jones** **Duration 1.03**

[Inset Video/Elections/Jones/Five]

In Words: [Sound Up] And with that I declare (Cheering)

Out Words: ...time for this Government to go.

[Capgen at 3"]
[JOE JONES\Political Correspondent]

Figure 12.6 The accurate duration of every video or audio insert, together with the In Words and Out Words, are regarded as essential information to be included in a script page on the computer screen. The information on this page is for the transmission of the video report on the election scenes and would follow immediately after the presenter has read the script in Figure 12.5. The reporter's name appears as a word Caption (also known as Aston or CG). Conventionally a NAME is in capital letters and the Title (or what a person is) is in normal title case.

Five O'Clock Report **Elections/Live Link** **Duration 1.30**

[Inset Read + Inset]
[John In Vision]
Opposition delight is to be expected. But where does this really leave the government now? Our political correspondent Joe Jones is at Westminster to give us this assessment. Joe – is the government genuinely in trouble or is this just a setback they can easily overcome?

[OS next]

[Live Event: JOE JONES X LIVE LINK/WESTMINSTER]
[Capgen: JOE JONES\Political Correspondent]

Figure 12.7 A script page must be put into the system for every item in the running order, even though the details might be sketchy. This is for a live contribution by the Political Correspondent – a two-way summary in which the presenter 'links' to an Outside Source (OS) where the Correspondent is standing waiting to respond to a few questions. Because it is live, the writer is only estimating that the news item will be 1.30 in duration. A Capgen is available to the director but because this is live the time for it to appear is not set on the script. This is because the director may want the freedom to bring it up when it seems appropriate.

The individualistic approach to script layout is matched if not exceeded by the way the terminology of television is used. As well as differing from one news service to another it is not uncommon for it to do so between stations within the same organization.

For instance some stations are happy to use 'videotape' and 'caption' as generic terms, without bothering to differentiate between formats and styles, or to refer to items of equipment by their trade names. While in the control room during one newscast the command 'run VT' is accepted shorthand for 'put videotape product made by the Sony Corporation on the screen', in others 'run/take Vt/cut to report/play it!' will mean the same thing.

Television has defied several brave attempts to establish a definitive glossary which would iron out all these inconsistencies. Compatible computer processing and output systems have helped to eliminate a lot of the differences in language used. Maybe it scarcely matters anyway. Uniformity, however desirable, is less important than ensuring everyone concerned in the production of news understands and reacts correctly to whatever local terminology is considered acceptable.

13

Production and presentation

As tension rises among the editorial staff in the newsroom, at the same time the focus of the entire news operation begins to shift inexorably towards the production and technical teams who will shortly be involved in turning all that journalistic effort into a television broadcast. The keys to this can be found in the studio and its adjacent output suite or control room. Computer terminals link everything. The news journalists must always remind themselves that computers do not make mistakes. They are only tools. Only people make mistakes – even if that includes the people who programmed the computer software in the first place!

The control room

Not unlike the bridge of an oil supertanker or the cockpit of a transatlantic jet, the control room of a television studio is a world of its own, dominated by wafer-thin monitors and electronic gadgetry of seemingly overwhelming complexity and luminosity. The main source of light comes from a bank of screens which reflect the seeing eyes of the cameras in the brightly-lit studio nearby. Other small squares of light shine from the illuminated buttons on control panels, at which sit shadowy figures, heads bent over news system screens. Tension and excitement lie close beneath a surface air of calm efficiency.

This, then, is the nerve centre of a television news broadcast, the one and only place where the editorial function, supreme until now, has to take second place. For this is the moment when the production and technical teams hold sway as they set about the intricacies of translating plans into something resembling a living television programme. No matter what the number of sources – live, recorded, visual, oral, static, moving – they have to knit the whole structure together, each dovetailing neatly into the next to produce a continuous, seamless whole, based on split-second timing. No matter what chaos explodes, it is what comes out of the TV sets of the nation, or the world, that matters.

That is not overstating the case. Split seconds make all the difference between a programme which flows from item to item without a hiccup and one which is untidy and ragged, with awkward delays between presenter introduction and the start of a report, clipped sound, momentarily blank screens or missed cues. Yet there is a very slim margin indeed between success and failure. The most ambitious programmes,

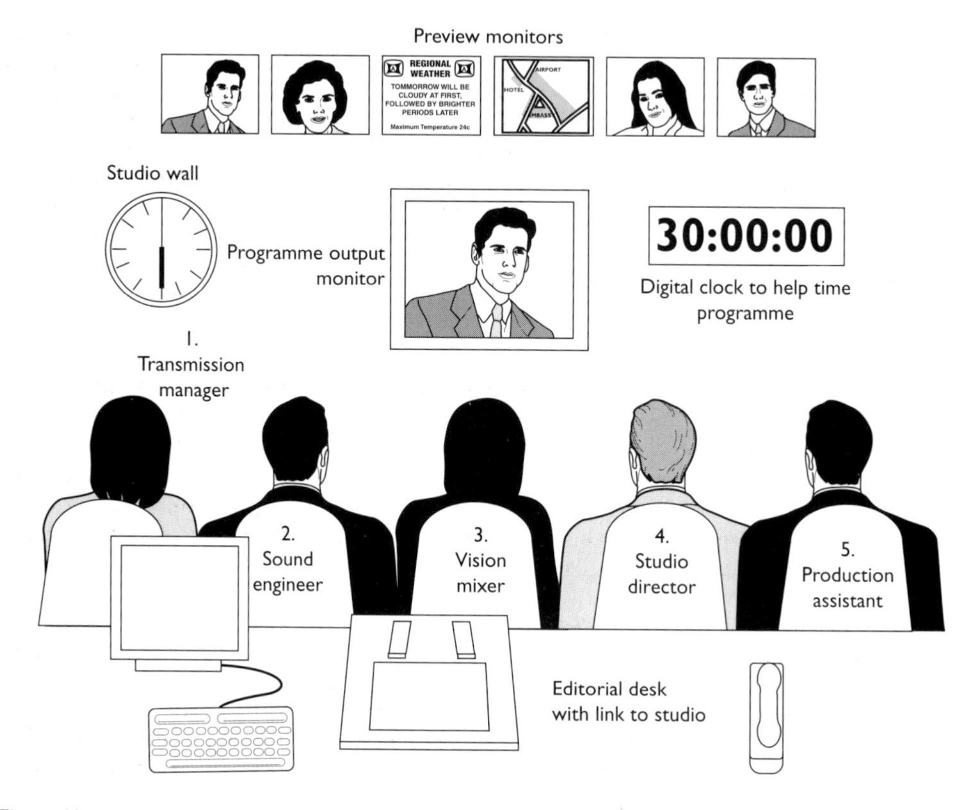

Figure 13.1

especially, court disaster day after day, pushing men, women and technology to the limits by the deliberate policy of trying to squeeze in the very latest information. New stories are added, some are dropped and others altered right up to and including the time the news is on the air.

As a result, the professionalism and expertise of the control room staff is continuously on trial. One slip, and all the time, effort and money spent by others might be wasted. Fortunately, real disasters are rare, despite the knowledge that one misfortune has the habit of begetting another, and the prickly feeling that, one day, an entire programme is going to collapse on air like a house made of cards.

Responsibility during transmission lies as much with the director as the journalist acting as editor. Back in the days when news bulletins came in short bursts of 10–12 minutes it was reasonable to expect directors to be responsible for several during a single-duty period. Single programmes and continuous news channels have become technically more complicated. Proper preparation is essential, and a director is now more likely to be associated with the same programme every day, at least on a short-term basis. The aim is to provide continuing production advice to the journalistic team on the ways items might be prepared for the screen and to help maintain the overall continuity in presentation which gives the programme its identity. In modern digital-based news services, which are often testing new ways of working with technology that is constantly evolving, a single technical director can have the facility to do jobs that were previously carried by two or three people.

As ever, the extent of any directoral duty depends on whether he or she is 'dedicated' to one news programme or to the news service as a whole. In a less-than-ideal world the director may come cold to a programme shortly before transmission, having already been part of a team involved in other output. In quarters where the journalists have a reputation for being difficult or the news department regarded as the poor relation, unworthy of creative attention, directors may be forgiven for believing they have transgressed in some obscure way and have been rostered to work on the news only as a punishment. In return, the editorial team is likely to greet the efforts and commitment of the director with suspicion and resentment. It is scarcely surprising if programmes broadcast in this atmosphere of mutual animosity show up on screen as dull and unimaginative.

Far better results are usually obtained from the director's consistent and direct involvement with a programme. Attending the editorial conferences, organizing and supervising graphics sequences or studio interviews are virtually certain to guarantee a familiarity with the content which will enhance production values. Some directors help select headline sequences. The overriding concern is to keep a close watch on the running order, to be ready to sound the warning bells over any potential difficulty, and to keep abreast of changes as they are made. In short, the directors of the most successful programmes are those who act as the focal point for everything to do with its production.

The design and positioning of control rooms are influenced by circumstance and fashion. A trend which has seen programmes broadcast direct from newsrooms rather than from separate studios makes it sensible to construct transmission facilities in a designated area nearby, but perhaps more typically the production team are housed in a separate gallery overlooking the studio or in a room next to it. The director will probably be seated more or less in the centre of a desk running virtually the whole width, using 'talkback' to communicate with everyone.

Journalist PC desktop → Editor PC Desktop →

→ Media Manager + Digital Output + Editing System →

→ Technical Director → Transmission

Figure 13.2 Speeding up transmission. Automated computers, centrally controlled, allow digital sounds and images for separate news stories to be selected from codes put into the journalists' PCs.

The director will operate from a computer screen which has the same information inside it provided by the programme editor: running order, scripts and technical instructions. The director has to turn this into real-time television. Banks of sliding switches and glowing buttons represent the selection of cameras and other sources available to be put on the screen by the director. In bigger news output areas the director may have a vision mixer, sound engineer and a production assistant to help him, but in modern digital-based suites he may have the facilities to do it all himself.

These specialists, then, form the basic studio control room production and technical team, although there might be an additional operator controlling the robotic cameras which move about without human guidance to pre-programmed positions like a mechanical dance.

A place in the control room is usually allocated for the programme editor and perhaps one of his or her colleagues as backup. With all these experts on the spot, answers to any problem which may crop up during transmission can be supplied very quickly.

The studio

From about 1980 news was often presented directly from newsrooms with a backdrop that resembled a computer warehouse. News managers thought the public would be impressed by the high-tech look. Designers advised them that a backdrop of computer newsrooms looked chic and efficient. By 1995, when everyone knew what a computer looked like and was no longer impressed, separate studios made a comeback. Then by 2000 television news was back in the warehouse, with a busy newsroom or monitor bank background – this time, because it looked less formal and relaxed. It was also cheap.

The new stage of news studio development is the virtual reality studio where only the news anchor/presenter is real. The problem news managers worry about is that it tends to make the news look like someone with a constant fidget. If the background can be adjusted in less than a second then it is tempting to do just that. The emphasis of the VR studio is simplicity and familiarity. The irony is that the technology can make any change the news organization wants within seconds, but the viewers may not want the look of the news to be persistently changed.

Figure 13.3 The gallery of a television studio ready for transmission. It is like the flight deck of an airliner. Computers control all systems but only people can control computers. Banks of monitors showing location links illuminate faces in the gallery and studio shots. The experienced director will try to cultivate calm in the one single minute before live transmission, no matter what chaos has gone before. (Photo courtesy of and © Sky Television.)

Big television productions still need plenty of space and technical support. Studios exclusively for news tend to be small and basic, the layout and contents often fixed for the long periods between revamps. A close look at almost any news programme will show how little of it is spent in the studio compared with the time devoted to routing video and other sources through it. Even so, as it is largely on the presentation of studio-based items that programme identity is maintained, considerable thought has to go into their production. To do that effectively might require three cameras. One concentrates on a head and shoulders shot of the presenter, one on a wide shot of the whole studio and a third reserved for additional contributors – reporters, correspondents and other interviewees.

Movements are so few and unfussy that some organizations find it just as effective to use the robotic cameras operated from the control room. Each camera is equipped with an electronic memory capable of storing pre-selected shots for use on transmission, thus removing much of the mental strain from the camera-work needed in a fast-moving programme.

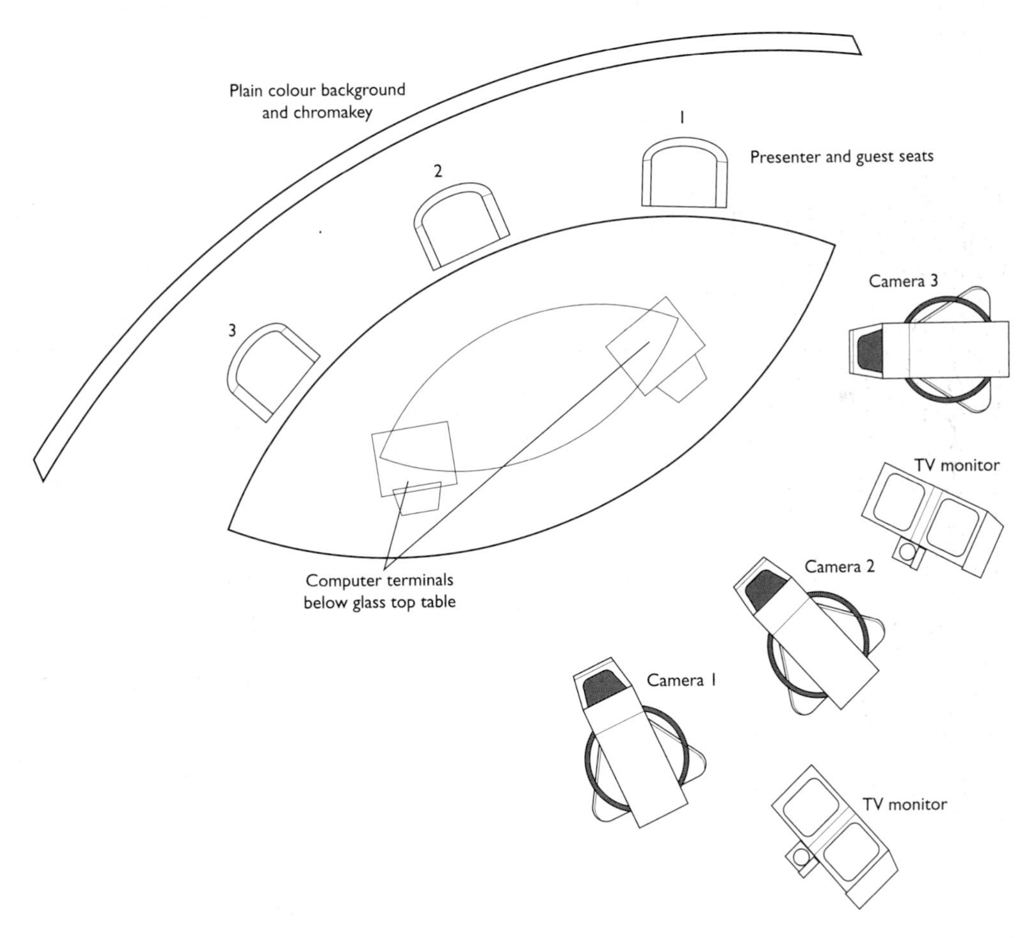

Figure 13.4 Common camera layout for a news studio. The 'high-tech' look of modern sets is often created by clever use of lighting, electronically generated backgrounds and computerized 'virtual reality' images.

Sound and lighting

Studio sound is fed independently from microphones – large or small, in or out of shot according to taste – which are usually sited in pairs, in case one fails, at each of the reading positions.

Overhead, banks of individually adjustable lights hang down from hoists close to the ceiling, throwing their beams over studio sets designed to enhance the special qualities of the programme. Every customer programme passing through the same studio will have its own needs. 'One-off' productions require special attention, but for the frequent occupants new technology is probably on hand to help, here as everywhere. Computer programs ensure optimum lighting plots can be repeated accurately from the control room with the minimum of delay. Lights in studios dedicated to one or a series of similar newscasts will probably need maintenance and only minor adjustments on a regular basis. The main necessity is to ensure even light levels across the set and to reduce the occasions when presenters and contributors appear in unflattering shadow.

Sets

The considerable thought and expense which goes into the construction of studio sets for television news hardly seems commensurate with what at bottom amounts to one basic requirement: somewhere for the cast to sit facing the cameras. Yet all news organizations throughout the world constantly revamp their sets, not just because of the need to demonstrate change and innovation but to give their product a fresher look. Cool colours go out. Warm colours come in. Clinical steel gives way to warm wood or cool glass. Straight lines give way to curves. Minimalism today becomes multicoloured variety tomorrow. It is in the end not much different from anyone's private home. The sofa you bought a few years ago was once admired, but now looks like it needs a new cover.

The BBC's television news programmes once underwent a complete refit and included the introduction of virtual reality: a computer-generated 3D image created the impression that much of the studio was occupied by the Corporation symbol, etched on a huge piece of glass, which disappeared as the opening story was read. One effect was to make the studio seem bigger and more imposing, especially in the opening and closing sequences which gave the impression of a long shot. Yet practically nothing was real apart from the presenter sitting at a minimal desk equipped with computer. For the designers involved, it emphasized their chief dilemma: how, at a time when the range of news channels and programmes was growing, to establish a unique and interesting 'look' the audience would identify immediately without being distracted. The BBC, ITN and Sky all moved away from cold colour schemes to warmer shades: wood panels, terracotta and cream, soft and comfortable shades. The space-ship look was dumped in favour of the Victorian parlour.

Some programmes are content to have a discreet logo on the screen at all times, but it is still possible to find fixed or removable backings made of wood or plastic, and the fashion which once saw newscasters working in studios surrounded by real or simulated bookshelves may also live on somewhere. Placing presenters against 'windows' which allow the viewer to look out on familiar views continues to be popular, while some programmes make use of real windows overlooking city landscapes, perhaps with a visible and carefully framed clocktower to add an urgent sense of passing time.

But probably the most popular layout for a modern news studio consists of a desk or desks placed against a plain background. For some years now it has been the vogue to

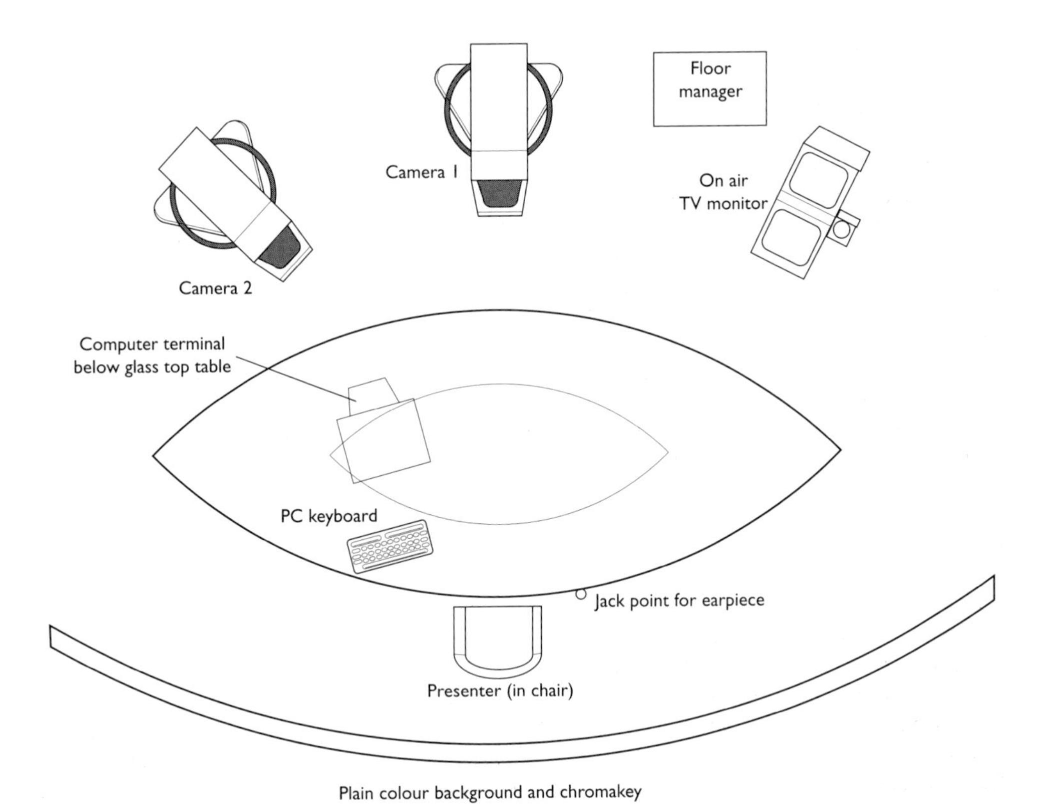

Figure 13.5 Smaller news studios may make do with two cameras and they are often robotic. This means they can move across the studio floor without a camera operator and instead can be positioned by a director from a gallery. A variety of shots can be stored and selected. For safety in the studio a Floor Manager is usually present.

fill part of this with a visual representation of the item being introduced, and it is quite common to see a still picture, a frozen or moving frame of video, or – increasingly – a specially designed graphic positioned over the reader's shoulder.

With careful thought and used sparingly, these devices, often known by the term 'inset', can be helpful signposts for audiences viewing multi-subject newscasts. This is often created by an electronic process known as chromakey or colour separation overlay (CSO). The secret lies in the colour chosen for the studio backcloth, frequently blue, of which there is very little in human flesh tones. At the touch of a switch in the studio the electronic camera locks on to it, filling it with a picture coming from a second source, perhaps another studio, a camera, stills store or video, effectively merging the two images into one. What is created for the viewer is an electronic illusion. The presenter in the foreground sees nothing except the plain studio backcloth, which can sometimes make two-way interviewing a tricky matter while this device is being used.

At one time when chromakey was used the foreground subject would occasionally take on a curious, blue-tinged outline, particularly round the hair, and zooming movements by the camera had the effect of making the foreground leap forward by itself. Most of the minor flaws in the system have been overcome, but performers still need to select clothing colours for the studio with considerable care, as the device is extremely sensitive. It will be triggered by any strong foreground colour which matches the background, punching an 'electronic' hole through anything in the way, human bodies included, and replacing it with the background picture.

Apart from the desks and the cameras, the rest of the floor space in the studio is taken up by coils of cables and portable monitors which enable the presenters to keep a discreet eye on both themselves and the output. Then there is the problem of communication between programme editor and presenter. Some programme editors like to be able to give instructions directly into their presenters' earpieces during transmission and are equipped with a link which is activated by the touch of a button in the gallery.

Script conference

An hour or so before a newscast (the exact timing depends on the duration and complexity of each programme) the production team assemble in the newsroom for a script conference. At this stage in the proceedings the director has had the running order for some time, at least some idea of the alterations which have inevitably been made since it was compiled, and advance warning of what other changes may occur before the programme goes on air.

Now the director is able to begin issuing detailed instructions, referring to the items on the running order in sequence, page by page. Everyone with a part to play in the newscast has to become familiar with the camera shots, the machine-by-machine allocation of videotape inserts, and the order of graphics and stills, for which a separate technical running order might have been prepared. Instructions are marked on whatever pages of script have been distributed at that time. Whatever details are known about pages yet to come are written on a skeleton, a sheet which is left completely blank apart from the corresponding page number of the script. When the completed page eventually turns up, the skeleton will be discarded, although as a last resort the director could work from the blank.

If we were able to eavesdrop 45 minutes before *The Five O'Clock Report* is scheduled to be broadcast, we would hear the dialogue for part of the script conference go something like this:

> *(Director speaking)* '... so it's page 22 next – Damages. That'll be camera two with an inset of the car the woman driver was injured in, and a map of where the accident happened. That's followed by page 23, Damages Interview – speaks for itself, I think – video on Line Two. OK? From there we go to the OB outside the court. It isn't in position yet, but should be within the next five minutes or so. Mustn't forget the "live" super this time. Page 25, Damages Background, is still in for now – video on Line One – but I'm told that's a very strong candidate for the drop. If it does go we'll still keep page 26, Damages Trail – and the car inset again. Hang on ... by rights we should be going back to Camera Two: I'll check. Right. Nothing then until page 30, Halfway. Camera One. Pages 31, 32 and 33 are the video inserts, which I'd like to record after the script conference if we get the chance, otherwise we'll keep them separate. Line Two for Farming, One for Orphans and Two for the Cricket Record in that order. Page 34, Crime Figures. Camera Two and inset of the report, into page 35, Crime North West, video on Line One and super of the reporter's name. Now a possible big change on page 37, Crime Chief. They're actually hoping for a live interview. John will be doing it down the line if it happens, but that'll have to be confirmed ...'

The conference continues briskly in the same vein, with pauses for the director to answer queries from other members of the production staff. Eventually every page on the running order has been covered, and allowance made for those items about which very little is known so far. Everybody concerned with the production is now fully aware of the part he or she is expected to play, while being quite prepared to accommodate any sudden change of plan. It is a fairly routine day for news, however, and with the favoured editorial shift on duty, the production and technical teams are hoping for a smooth, good-looking programme.

Rehearsal and transmission

Rehearsals for television news programmes tend to be sketchy affairs because so much of the material to be used on transmission is unavailable until the last few minutes before airtime. Scripts are still being written, video edited and graphics completed. In fact the assembly of items goes on so late that programme editors are often heard to grumble that they scarcely have any time to see some of them to make editorial judgements, let alone expect the luxury of being able to rehearse.

Just the same, the director must take the opportunity to go through whatever material is ready, and the one responsible for our *Five O'Clock Report* is no different. Until now the northern report on the crime figures and Delia Ward's Topfield Farm item have been seen only by the editorial staff, so the availability of both means the production team can have a preview. A halfway headline sequence is also ready: the script is still being written back in the newsroom. At least the script to accompany pictures of the post-by-election scenes is finished and the presenter canters through it to see whether the words will match. The live link between the studio and the unit outside the 'Damages' court has been established by now, although all the camera shows is a shot of a paving stone covered with television cables.

As the minutes tick away the flow of completed scripts coming into the control room begins to quicken. The corresponding skeletons are thrown out to accommodate the new pages, everybody remembering to transfer the marks they made so carefully at the script conference earlier.

Editors who for no apparent reason make wholesale changes to their programme running orders at the last moment are not popular with even the most experienced production teams, and acquire reputations for being indecisive. Inevitably, though, the duration of some items will turn out to be longer than planned and others will have to be adjusted or omitted to fit them in. Minor alterations are unavoidable.

Director: 'Page 34, everyone. There's an extra chart on graphics. All right?'

Sometimes there are deletions, ordered by the editor sending information from the newsroom on the news computer message system.

Editor: 'Don't want the Damages Background, thanks.'
Director: 'Just to confirm: page 25, Damages Background, is dropped. That affects you, Line One, thank you.'

In between Michael, the economics correspondent, wanders into the studio and is motioned into a seat opposite camera three. The director's voice booms into the control room.

> *Director:* 'Can you try the voice level for Michael now, please? He's got to get back to check up on a couple of things before the programme. Sound?' *(He presses a button on the desk so the correspondent in the studio can hear.)* 'Just a few words then, Michael.' *(After no more than a sentence the sound engineer in the control room has adjusted the fader and given the thumbs up.)* 'OK. Thanks.'

With not much more than 15 minutes to go, the editor in the newsroom is confronted with the sort of dilemma he has nightmares about. The regional reporter covering the joyriders' story has called the editor to say four people have been killed and seven, possibly eight more seriously injured in the traffic pile-up on the motorway. This has only become known since rescue services got through the chaos and found the dead and injured in two cars and a mini-bus in a ditch. He's offering an interview he has just carried out with a local police spokesman, and separate, unedited rushes of activity at ground level, shot by another crew. He hasn't seen these pictures, but understands they contain unpleasant detail. To cap it all, it seems the injured joyriders had absconded from a juvenile detention centre two days earlier, but no one other than the local police had been made aware of it.

The editor's instinct tells him this has all the makings of a serious issue. By themselves, deaths on the motorway – regrettable though they are – do not always cause public outrage. The circumstances of this accident are entirely different and the ramifications considerable. There is still time to switch the lead, but it would mean wholesale changes to the running order and destroying much of the planned structure of the programme. Other thoughts occur: there are obvious links with the 'damages' story lower down.

How did the joyriders abscond? Was there a security lapse? Have official guidelines been breached? A specialist correspondent should be finding out which government department is responsible for policy.

The editor also wonders whether there is time to throw together a brief background on current methods of dealing with juvenile offenders. One more question needs to be resolved:

> *Editor (to the newsroom team):* 'What's the word on our report author – would she make it into the top half of the programme?'
> *Newsroom (looking at the newsroom clock):* 'Should be on the doorstep any minute now.'

The editor makes up his mind.

> *Editor (to his no. 2):* 'I'm going to switch the lead. We'll move the joyriders up to page two with the intro, map and helicopter videotape, and follow it with the ground level pictures. I'll see if we can get the local reporter's voice over them. A reporter is chasing someone from the detention centre where these boys were supposed to have been held. If they don't manage it in time for page three we'll try and slot them in later. I'm also going to move up the crime story. There isn't room for it all on page four ... Just warn the control room what's likely to happen. I'm just going to have another word with the newsroom, then I'll go and talk to the director. In the meantime, you'd better get the headlines rewritten. We'll drop the Crime North West now.' *(He consults the running order.)*

In the semi-darkness of the control room along the corridor, the director is still in the process of catching up with earlier changes.

Director: 'There's a new story, Transfer, which they're calling page 47a, after the Picasso and before the cricket. Camera two in close-up for Val, plus a still of Jamie Jamieson. And page 35, Crime North West, is out, so we won't need the VT on Line One, thank you. Tell John it means we go straight to his down-the-line interview with the report author. Camera Two, I think we'd better make that a close-up on page 36 then. OK? Right. Script check in five minutes everybody ...'

By now the final scripts are being rushed in. Everybody is writing furiously into the computers, adjusting scripts and technical instructions. The main presenter is reading the lead story quietly to himself. Everyone is complaining about problems with the way script and technical instructions have been put in. When it gets tense every mouse is being treated like a hammer pushing in a nail.

The editor has arrived in the control room, out of breath after confirming that the regional reporter covering the joyriders' story has sent a separate voice track to cover the pictures of the helicopter and the scenes on the ground. These are now being edited together and should be ready in no more than five minutes' time. The shift reporter is trying for the head of the detention centre, which by chance is in Oxfordshire, forty miles from Topfield Farm. The camera crew and links vehicle which covered that story have been diverted and should be there before the end of the programme. So far, though, the government department responsible for juvenile detention centres hasn't agreed to let the chief executive talk.

The director listens, unperturbed, to the catalogue of changes being made. He has been through all this, and worse, many times before. Now he speaks calmly into the talkback microphone.

Director: 'Change of lead coming up. There are knock-on effects, too. Tell you about them all in a moment, during the script check.'

The writer responsible for the lead story rushes into the gallery. 'You'd better come and have a look at the crash pictures,' he tells the editor. 'They're pretty gruesome and I'm not sure what we should show.'

'I'll have to leave it to you,' says the editor. 'We're about to change the running order. Just make sure there are no lingering close-ups.' The editor is telling his journalist to use his initiative. 'Do my best,' says the writer. 'But there's a lot and it'll take time to edit them out. I hope we'll be ready.'

The final changes are written into the running order. The editor is still busily using the computer to do his calculations. The joyriders' crash, the damages story and the crime figures sequence have suddenly made the programme very top-heavy with crime and cars, more so if he has to drop material to make way for another interview. He's now a good three minutes overset. The easy way would be to drop Topfield Farm, which alone would save three minutes. He could keep the European element, which is a story in its own right. The trouble is he doesn't relish facing Delia Ward afterwards. He looks around the control room, seeking inspiration, and notices the preview monitor on the stills store, which is just lining up on its subject of the Transfer story.

Editor: 'Is that the Jamie Jamieson still? I thought he played for Barcelona. That's a Juventus shirt.'

The stills assistant hurries off to check with the writer back in the newsroom. Five minutes to go. 'Script check,' orders the director. This is the final check before transmission, and everyone is expected to pay full attention. Loss of concentration here could result in disaster later on. Again, the director is in command, going through the complete running order, page by page, in rapid confirmation of all that has gone on before and since the first script conference nearly an hour earlier:

> *Director:* 'Here we go then. We start with the *Five O'Clock* titles and signature tune on Cart One. Then there's a rewrite of page one, Heads, on its way. That's John's voice over Cart Two of the motorway, Graphics One of the Crime Figures report, Cart One of the election scenes and Cart Two of the cricketer. Into the new lead, Joyriders, which is now renumbered page two. John and inset of the crash, map on Graphics Two, vt package on Cart Two, which is going to be very very late. Page four: Crime Figures. Val on Camera Two with an inset of the report, charts on Graphics One and then straight into a down the line with the author of the report. To confirm, it's Val doing the interview, not John as originally planned ...' *(To the editor)* 'Your author's going to make it in time, is she?'
> *Editor:* 'Definitely.' *(No one has dared tell him the regional studio in which the author is due to appear will be occupied by a children's programme until approximately 90 seconds before the news begins.)*
> *Director:* '... then fourteen, Election Political, the OB in Downing Street. Mustn't forget the "live" caption. Don will end with a handback to the studio. John picks up on page fifteen, the economy ...'

The director continues to rattle through it all, remembering to remind everyone page 25, Damages Background, is out; Crime Figures, originally between pages 34 and 37, has been moved and renumbered; and Joyriders, previously at page 40 and 41, is now at page two. The few queries take only a few seconds to answer.

Red flashing lights outside the studio and control-room doors indicate impending transmission. The studio begins to settle down. The prompter operator is at the keyboard of the computer which allows the presenters to read their scripts while looking directly into the cameras. Val glances down at the notes she scribbled during a quick conversation with John, hoping she will be able to decipher them when the time comes to carry out her interview. Music filters into the studio, signifying the approaching end of the preceding programme. (Pictures from the regional studio where the report author should be appear in the control room on a monitor to the director's left. Several teams of children are engrossed in some elaborate quiz game.)

One minute. A writer races into the control room with the rewritten headline page. The known death toll in the joyriders' crash has now gone up to eight. At the back of the control room a fierce argument is going on about the Jamie Jamieson picture. Voices rise.

Thirty seconds. The director reluctantly tears his attention from the monitor bank.

> *Director:* 'Quiet please! Stand by everyone.'

Station announcements are coming to an end in the brief break between programmes. A square dot appears in the top right-hand corner of the screen, which is now showing station identification and a clock.

> Fifteen seconds. The dot disappears.
> Ten seconds. Nine. Eight. Seven ...
> *Announcer:* '... that's at seven-thirty. But now, at five o'clock ...'

Director: 'Stand by Titles.'
Announcer: '... it's time for the news.'
Production assistant: '... four, three ...'
Director: 'Run vt. Coming to studio and voice over ...'
Production assistant: 'Counting into opening headline ... two ... one ...'

The titles and signature tune come to an end.

Director: 'Cue him.'
Presenter: 'Eight people are dead and four seriously injured ...'
Director: 'Graphics One next ...'
Production assistant: '... counting out of vt – four, three ...'
Presenter: '... after joyriders crash on a motorway ...'
Director: 'Cut. Standby Vt One.'
Presenter: 'Unemployment among the young ...'
Director: 'Run vt.'
Production assistant: '... two, one ...'
Presenter: 'After last night's two by-election defeats ...'
Lead story writer (to editor): 'All done – the vt's ready. And not too much blood.'
Editor: 'Better not be. Any sign of our author lady? We'll need her in a couple of minutes.'
Director: 'We'll be coming to Camera One for page two. I said one!'
Production assistant: 'Thirty seconds, Vt One.' *(Buzz)*
Duty transmission manager: 'Can't see her.' *(He indicates the preview monitor. The children have gone, but the studio is now ominously empty.)*

The director, following his script, waits until the presenter in the studio gets to within nine words of the end of the introduction to the joyriders' story. Then:

Director: 'Run vt!'
Production assistant: '... three, two ...'
Presenter: '... just received this report from Tom Dixon.'
Director: 'Cut to vt.' *(He relaxes)* 'A little more headroom on your presenter shot, please Camera Two. She's looking a bit squashed. Has our interviewee turned up yet?'
Duty transmission manager: 'Doesn't look like it.'
Director (grumpily): 'What am I to do if she doesn't make it. Show an empty chair or go on to the next story? We've got under two minutes.' *(Wearily)* 'Will they never learn?'
Editor (reassuringly): 'She'll be there.'
Production assistant: 'Thirty seconds left on this vt.'
Director: 'Standby studio. Standby Camera Two.' *(To editor:)* 'Are you happy with the Jamie Jamieson still yet?'
Editor: 'We're working on it.'
Production assistant: 'Ten to go.'
Director (seeing movement on the monitor showing the regional studio): 'Aha ... methinks I detect signs of life out there. Coming to Camera Two ...'
Editor (sounding smug as he wipes sweat-damp hands on his handkerchief): 'Told you she'd be there, didn't I?'

Now the interviewee has arrived it seems certain Topfield Farm will have to go after all. It will mean a change to the halfway heads. Delia Ward will need a big drink after the programme. So will he.

Production assistant: '... three, two, one ...'
Director: 'Cue her and cut!'

Although they are unlikely to admit it in public, hardened newspeople still manage to marvel at the way production staff throw an almost entirely unrehearsed programme on the air without so much as a tiny hitch. The occasions when minor errors do occur – the director running a videotape too soon, the vision mixer pressing the wrong button, the sound operator fading up the words of an interview a second late, the graphics designer misspelling the place name on a map – are usually the subject of lengthy and heated inquests. Audiences probably do not notice until the mistakes are glaring, and then they seem to take a perverse pleasure from the knowledge that their favourite news programme is peopled by fallible humans like themselves. Perhaps that accounts for the enormous popularity of those television programmes composed entirely of scenes which have gone wrong.

There is a serious side to it. The reputation of television news rests above all on its editorial credibility (witness all that mail and those telephone calls from viewers) and if programmes become so riddled with production errors as to make technical 'cleanliness' impossible to maintain, there is a real prospect of the journalism itself eventually becoming undermined.

Most of the mistakes which do occur could be avoided, but that would mean setting strict deadlines to ensure full rehearsals. It is not a realistic proposition for most news programmes. Television newspeople are acutely aware of the 'now or never' nature of their work. That is why editors are prepared to jeopardize an entire production for the sake of including a late story breaking halfway through a programme. By dropping one on-camera item, taking the option of an 'early out' on a video tape insert and striking out all but one of the closing headlines, there is suddenly enough room to squeeze in something which may already be too late for most editions of tomorrow morning's newspapers. And that, television journalists will say with satisfaction, is a large part of what it's all about.

Local and regional news

What has gone before has been principally to do with programmes which bring news to national audiences. But these programmes, despite their prestige and the large audiences they command, are easily outstripped in total numbers and popularity by those concentrating on regional or local matters. At its best, regional television journalism is able to offer a more accurate mirror of society than its network counterpart can ever hope to reflect, and sensible politicians anxious to keep a finger on the public pulse often consider the programmes essential to see and be seen on.

Content, in programmes sometimes longer in duration than the 'national' news, is often a broad mixture of the serious and lighter issues, the contentious matters either purely local or national ones given a local twist. The 'news' section, sometimes read by a separate presenter, comes in the shape of mini-bulletins containing graphics and fast moving video inserts. On the 'magazine' side the daily reporter packages and studio discussions are supplemented by regular feature items which might include consumer matters, comprehensive sport, arts and entertainment news. The trick with programmes

RUNNING ORDER. FIVE O'CLOCK REPORT.

no.	title	source	writer		duration	cum. dur.	checked
001	TITLES/HEADS	VTS	JB	R	0.45		ap
002	JOYRIDERS	CAM 1/INSET/MAP/VT/ SUPERS	FS		2.45	03.30	ap
003							
004A	CRIME FIGURES	CAM 2/INSET/CHARTS	MR	R	0.50	04.20	ap
004B	CRIME/CHIEF	2-WAY/SUPER			2.00	06.20	ap
005	ELECTION/INTRO	CAM 1/INSET/STILL			0.48	07.08	ap
006	ELECTION/SCENES	VT/SUPERS	CR		1.03	08.11	ap
007	ELECTION/REACTIONS	CAM 1/INSET			0.32	08.43	ap
008	ELECTION/INTVWS	VT/SUPERS			1.30	10.13	ap
009							
010							
011							
012	ELECTION/REST	CAM 2/CU/CHARTS			0.48	11.01	ap
013	ELECTION/ROUNDUP	CAM 2/CU			0.18	11.19	ap
014	ELECTION/POLITICAL	OB/SUPERS			2.15	13.34	ap
015	ECONOMY	CAM 1/INSET	JL		0.15	13.49	ap
016	ECONOMY/JONES	CAM 3/CHART					
017	ECONOMY/WRAP	VT/SUPERS			2.53	16.42	ap
018	HALFWAY	CAM 1/CU	TC		0.08	16.50	ap
019							
020	ORPHANS	VT			0.10	1700	ap
021	RECORD	VT	RB		0.10	17.10	ap
022	DAMAGES	CAM 2/INSET/MAP	RS		0.40	17.50	ap
023	DAMAGES/INTERVIEW	VT/SUPERS			1.02	18.52	ap
024	DAMAGES/2-WAY	OB/SUPERS			1.37	20.29	ap
025							
026	DAMAGES/TRAIL	CAM 1/INSET	TC		0.15	20.44	ap
027							
028							
029							
030							
031							
032							
033							
034							
035							
036							
037							
038	KIDS	CAM 1/CU	RA		0.18	21.02	ap
039	KIDS/SMITH	VT/CHARTS/SUPERS			2.21	23.23	ap
040							
041	JOYRIDERS/LATE	CAM 2/INSET			0.19	23.42	ap
042	JOYRIDERS/HEAD	VTY/SUPER	FS		2.12	25.54	ap
043	TOPFIELD	CAM 1/CU	MR		0.20	26.14	ap
044	TOPFIELD/EUROPE	VT			0.32	26.46	ap
045							
046							
047	PICASSO	CAM 1/INSET			0.16	27.12	ap
047A	TRANSFER	CAM 2/STILL	PT		0.10	27.22	ap
048	CRICKET	CAM 2/INSET	JG		0.15	27.37	ap
049	CRICKET/AUSSIE	VT			0.55	28.32	ap
050							

Figure 13.6 The running order of *The Five O'Clock Report*, as broadcast after taking account of all the developments during the few hours leading to transmission. 'R' denotes a rewritten page. Now compare with Figures 12.2 and 12.4.

scheduled on the fringes of evening peak viewing time is to devise a combination appealing enough to stop viewers switching to another channel immediately afterwards.

The pace in a typical local newsroom is bound to be more leisurely, and some of the issues less immediate, but for the aspiring television journalist, work on a regional programme may be far more stimulating than one nationally-based. Staff numbers are invariably smaller and individuals harder pressed, opening the way for those with ambition to turn their hands to many different tasks. As a result, a number who start out as local newswriters, producers or reporters soon find themselves being tempted by offers to join the bigger city or national programmes. In recent years, though, the professional satisfaction to be gained from working locally – to say nothing of the advantages of a better quality of life – has made it easier for many to resist, and the flow is by no means all one way

There is also something immensely satisfying about being so close to the grass roots. Many of the subjects handled by national newswriters and producers cannot be anything but remote from the vast audiences they serve. Local television journalists live with the knowledge that what they report and write about directly touches the lives of their viewers. It helps make them more careful. Instead of writing rude letters, complaining members of the public are liable to come round to the office and hammer on the door for an answer face-to-face, or accost any programme personality they recognize doing their weekend shopping.

Some local programmes are superbly resourced, with enough talent, equipment and facilities to put many a national newscast to shame. Modern links vehicles and satellite news-gathering has also given editors the opportunity to do more 'live' inserts, adding to the immediacy which, knowingly or not, viewers have now come to accept. Live links by reporters, linking in and out of videotape reports and following them up with live interviews, have become more common in both BBC and commercial television regional programmes.

For the editor of an under-financed programme, the biggest headache is usually how to fill an allotted 10, 20 or 60 minutes every afternoon. There may not be enough staff to allow for a continuously high level of planning, so much of each programme may have to be assembled from scratch a matter of hours before that day's transmission.

That leaves little room for manoeuvre, and should the already meagre ration of stories be reduced for some reason, the alternative to the unthinkable – leaving a hole – is to pad out what remains. The result is not attractive. The live studio interview may have to be stretched by an extra minute even though the subject has been exhausted long before. The 'news' slot may be so crammed with on-camera stories that it appears to be a convenient dustbin for dumping the oddments unwanted anywhere else instead of being a brief, crisply written round-up of matters of genuine interest.

The arrival of the lightweight digital camera has not provided a complete answer. The speed and flexibility it adds to coverage has, in some places, led to more trivialization of the news, with almost every minor traffic event, every fire, every petty crime hyped to a level of treatment it may not deserve, simply because pictures are available or easy to obtain quickly. In the edit suites, the picture editor may be encouraged to lengthen shots or include those better discarded.

Before long, if these conditions are allowed to persist, standards slip, fundamentals including proper shot-listing are ignored, and the whole programme becomes flabby and over-written. The journalist's ancient battle-cry of 'What's it worth?' is replaced by the anxious inquiry 'How long can you make it?' This may be an understandable attitude, but it is one which does disservice to the viewer. Almost every story has its own 'natural' length, in whatever context it may be found. Going beyond it does nothing except debase the coinage.

Headlines

The style of many news programmes now incorporates opening 'headlines', words and pictures representing in brief each of the main or most interesting stories, the purpose being to attract and keep viewers' attentions from the top of the newscast onwards.

As a rule these headlines are preceded or followed by a generic title sequence and stirring signature tune. Title variants include real or abstract impressions of the general geographical locations or scope of subjects covered by the programme and shots of members of its reporting team, and of the studio. The tempo of the music and design and intricacy of the title sequence might also govern the duration of the headlines, a restriction not necessarily appropriate for each occasion.

For all programmes, the headlines present the perfect outlet from which to 'set out their stall' for the viewing audience. Headlines work best with pictures with movement and action from the selected stories, preferably stories which will hold the audience through the entire programme. Editors who appreciate it make the task a priority, assigning their leading writers (or in some cases the presenters) to select the pictures and words with proper care.

In most cases it ought to be possible to identify potential headline stories early enough in the planning cycle to brief the reporters and camera teams assigned to them to be on the look-out for an especially telling picture or two. Many of the most effective are those headlines which have been shot specifically for the purpose and are not repeated in the body of a report. This does not mean losing out on the best pictures: a slightly different camera angle or framing is often quite enough.

Guidelines for headline pictures

- Be brief – one shot should suffice.
- Keep camera movement to a minimum – it can be distracting.
- People make the best pictures – buildings are boring.

Ideally, every headline phrase to accompany the pictures should be short, clear and strictly relevant. In five or six seconds – fifteen or eighteen words – it should be possible to get across the essence of most stories. Brevity is preferable to headlines which virtually tell everything, and clarity – one thought is better than two or more. Cleverly teasing headlines, however well written in their own way, may leave the viewer wondering ... and distracted.

Good headline writing can enliven even the least exciting day. As ever, of course, it is entirely subjective. For some tastes 'Eight British climbers were rescued after being trapped for a week ...' would not sound nearly as strong as 'Eight British climbers have been rescued after being trapped for a week ...' or as urgent as 'Eight British climbers are rescued after being trapped for a week', while 'Eight British climbers rescued: trapped for week' might be considered too much like a newspaper headline.

The present tense probably works best in many circumstances, adding an immediacy which no other can match, but headlines should be in the present tense only when the event is happening. Things that have already happened belong in the past tense.[1]

Therefore, even immediately after an explosion for example, one of the following would be correct:

1. *Style Guide* (2000). BBC News.

- 'A bomb explosion in the centre of Cape Town – three people are thought to have died.'
- 'The new South African Government has imposed immediate new laws of arrest and detention after a bomb exploded in central Cape Town.'
- 'This is the scene in Cape Town tonight as efforts continue to reach three people trapped following an explosion.'

Clearly in some cases the present tense is inappropriate – e.g. when referring to a time which is forthcoming:

- 'In the next hour, the government in Sierra Leone falls ...'
- 'In the next hour, as three bombs go off in Cape Town we ask – who still wants to stop South Africa's search for harmony and peace?'

How many headlines? There is no set rule. Logic suggests the longer the programme the more headlines can be accommodated, but avoid a lengthy list containing other than selections which represent the outstanding items. Three, perhaps four – five at most.

Insets and studio-set idents

A great deal of production effort often goes into creating the insets which appear over the presenter's shoulder in the studio in some news channels. Some combine graphics and video most imaginatively. The least effective are those which are too abstract for immediate identification by the viewer, or are out of proportion with the main subject – often a picture of a face – dominating rather than complementing the reader. At times, too, the writing does not always make a sufficient link to an inset, leaving the audience to guess the connection between what they are seeing and what they are being told. One reason for this could be lack of communication between writer and designer. There may not be a need in every case for a direct reference, but some positive link between the two ought to be established within the opening few seconds.

As with the headlines, the best illustrations are often those which have been designed especially for the purpose, although some designers – or more likely the producers who instruct them – sometimes seem to be uncertain whether their efforts speak for themselves and feel the need to add a written word or two by way of extra explanation. All but a few are irritating and unnecessary, coming across as patronizing to the viewers who may feel they are being told they lack the intelligence to understand a purely visual concept.

Halfway

No doubt the fashionable 'halfway headlines' or 'still-to-come' sequences will disappear as quickly as they arrived, but for the time being they represent useful devices for changing the pace of a programme or linking sections separated by a break for advertising.

'Halfway heads' present writers with a particular problem when all they offer the viewer is essentially no more than a repetition of the opening headlines. At this later stage of the proceedings, it could be argued, the emphasis ought to be different and room found for a subtle change in the words used perhaps only 15 minutes earlier to illustrate the top stories. The trick is to find another way of saying the same thing without losing the original impact in case members of the audience have missed the top of the programme. This is especially true in the early evening, when the switch-on factor is probably highest among viewers returning home from work.

'Still to come' offers two main promotional opportunities – for headline stories which do not make it into the first half of the newscast, and 'secondary' headlines for other attractive, perhaps less newsy items lower down the running order. But here words and pictures have to work hard to keep viewers watching, for the assumption is that a story which does not appear high up signifies its lack of importance.

End headlines

Closing sequences which include a recap of the main opening headlines, with or without pictures, ought not to be just a lazy duplicate of what has gone before. Running headline stories which have moved on should be updated and others reworded to change tense.

Linking to the weather

Where there is news, expect there to be weather. Or, to be more specific, a forecast transmitted before, during, or after a newscast. Weather conditions provide a fascinating talking point not confined to audiences in those countries in which information about climate is regarded more seriously than almost anything else. Is it a rule of nature that everyone is interested in the weather forecast.

For every television newscast the obvious importance of weather coverage as a service throws up a number of questions, the most crucial of which is whether it is part of and controlled by the news service, or an adjunct to and provided independently from it. The answer in many cases is both. Weather information, commonly placed around newscasts, may come within the remit of those who organize an entire programme schedule, while more localized forecasts are an integral and vital part of regional news.

Many years ago newscasts might have been content to let a presenter read out a short forecast as the last item – with or without an accompanying map or two – or hand the task to a not necessarily telegenic employee of the Met Office. Today, the daily weather slot lasting a minute or two is more likely to be one of the main ingredients by which the success of a news programme is measured. With it has come the creation of a breed of 'weather-casters', many of whom have become public figures in their own right, whether they are trained meteorologists or television professionals. The extent to which idiosyncratic weather-casters are allowed the freedom to dress or behave in ways which match the conditions they happen to be describing depends both on the style of the programme they serve and the editorial control exercised over them.

For the most part, though, newscasts benefit from offering forecasters who provide the weather outlook without too many frills. In either case the more scientific but rather dull recitation of information has tended to give way to easily assimilated summaries in a variety of styles and length against a background of maps and symbols from a separate studio or from within the same news setting.

Much of the raw data is provided by state-owned meteorological centres which gather it through a combination of satellites, ships, aircraft, rainfall stations and other sources, and have the scientists to analyse it. Increasing accuracy has allowed forecasts covering weekend and four-day periods to become a popular addition to programme weather-slots.

(a)

(b)

Figure 13.7 Weather forecasts, enhanced by advanced meteorological techniques and brought to life by computer graphics, are an essential ingredient of television programmes all over the world, sometimes as an integral part of the news, rarely far from it. (a) Sian Lloyd presenting ITV National UK weather on channel 3; (b) Andrea Mclean presenting *GMTV* morning weather for an outside broadcast in Australia. (Photos courtesy of International Weather Productions.)

Illustration of the weather has been revolutionized since the introduction of computer graphics systems (not all of which are as expensive as their sophistication might suggest). On-screen animation simulates the movement of weather fronts and cloud formations clearly and effectively.

With the improved presentation has come the need to introduce the weather more imaginatively. 'Finally, the weather' scarcely does it justice. Surveys by news organizations provided the evidence they needed to show that the weather forecast was a critical part of a news service. The on-air time period provided for the weather has grown considerably, and it probably deserves it.

14

Presenting the news

In the formative days of TV news the presenters mostly came from a theatrical background and were often treated as just another tool for the creative brains who put the news together. They had well-modulated voices and regular faces that did not distract the viewer. It all changed in the 1980s, when experienced journalists like Sir Trevor McDonald, Jeremy Paxman, Anna Ford and Michael Buerk moved into the presenter chair. Michael Buerk once said it was a very easy job. It certainly looks like big money easily made. Of course, it is not an easy job at all. Michael Buerk made it look easy because of his professionalism and skill. In fact news presenters – who fully understand the news they present – are really paid to be able to cope when things go wrong. An American President, Harry Truman, immortalized the saying: 'The buck stops here.' Conceivably, in his case, it was true. The widespread misconception about 'the buck' in television news is that it stops at the desk of the messenger seen to be delivering the message, good or bad, directly to the viewer. In short, the person known as the anchor, newscaster, newsreader or presenter.

How it began: newsreader or newscaster?

Fascination with those who undertake the role goes back well before the days when newsreaders changed employers for huge salaries. Newsreaders themselves have been known to compare their role with the town-criers of old, the main difference being that they passed on the word to the people from a comfortable seat in a television studio rather than from among jostling crowds in the market square. They also operated on a much more personal level, as the late Robert Dougall, who read the news for the BBC for more than 15 years, wrote in his memoirs:

> Television breeds a closeness and intimacy unlike that of any other medium. Your image is projected straight into people's homes. You become, as it were, a privileged guest at innumerable firesides. What is more, a newsreader is not playing a role, not appearing as another character, or in costume, but as himself. He therefore builds up over the years a kind of rapport with the public.[1]

1. Robert Dougall (1973). *In and Out of the Box*, Collins Harvill.

The truth of that was once gauged from opinion polls which consistently placed Walter Cronkite, a veteran of CBS News through most of the late twentieth-century, as being among the most trusted people in the United States.

The responsibility all this implied was fairly awesome. There was still more. A cough, hesitation or mispronunciation might easily make nonsense of the most serious or important piece of news. An erratic speed of delivery, especially during the few crucial seconds before the start of a news report, might have devastatingly destructive effects on the most carefully planned programme.

Yet between the early years, when the pioneers were expected to present the news anonymously off camera (in case an involuntarily raised eyebrow, facial twitch or some other expression could be construed as 'comment') and the beginning of the electronic revolution, the conventional newsreader was a rather contradictory figure. On the one hand he (and it was usually a male in those days) was accepted as the figurehead, the standard-bearer of the programme, admired and respected, the subject of unwavering public interest on and off the screen. On the other he was among the last to be consulted about content, style and format. Complaints that Reader A was prone to certain linguistic errors that Reader B scrupulously managed to avoid overlooked the fact that neither probably had very much to do with the way the words were written, only the way they were read.

So, sitting in front of a camera, reading aloud the fruits of other people's labour with the aid of a written script and an electronic prompting device, directed through a hidden earpiece by a studio director next door, supplied with a reference system to aid the pronunciation of difficult words or names, scarcely seemed exacting enough to warrant all the acclaim. No job for a grown-up, as one reader put it.

True, there were nightly butterflies to be conquered for those terrified of making a mistake or losing a place on the page in front of millions. And yes, the hot lights might make the half-hour or so of programme transmission a trifle uncomfortable – dangerous occasionally when, as has been known, a studio light exploded in a shower of glass – especially if formal dress were expected on set. But even here the 'standard newsreader shot' usually revealed no more than the upper half of the body. The rest might just as well have been covered in crumpled old jeans or a pair of shorts for all the viewers knew. Before Angela Rippon (a popular news presenter in the 1970s) danced on a BBC Christmas comedy show, critics of news programmes had been known to wonder, rather cruelly, whether newsreaders needed legs at all.

As for the qualifications necessary for the work, these would seem to have been limited to an authoritative screen 'presence', a pleasant appearance, clear diction, lack of irritating mannerisms, and ability to keep cool when things occasionally fell apart at the seams: talents that, it may be said, bore a striking resemblance to those required for 'ordinary' television reporting.

Not every newsreader was a reporter who had come in from the cold, and some reporters regarded with horror the prospect of swapping their passports for a permanent place in the studio. The original readers of the news were not necessarily fully fledged journalists at all, although they could scarcely have been effective unless they had exhibited a reasonable interest in the subject. Some began as actors, announcers for different kinds of programmes, or were simply chosen because they had the looks and voices to suit the fashion.

The outstanding quality the best of them managed to bring to television news was an almost tangible manifestation of the editorial credibility of the programme they were representing, allied to some superb skills in delivery and pronunciation which gave a de-luxe finish to writing frequently undeserving of it. For the viewer they became as familiar as old friends, welcome, immovable points of reference on a world map of constantly changing contours and values. No wonder one newsreader of those days is able to recall from memory virtually every word from letters written by lonely, elderly women who admitted kissing the screen every night as he ended the late headlines. Others must have their own stories to tell of receiving gifts, invitations, compliments, declarations of undying devotion, proposals of various sorts – and occasionally unwelcome personal attention – all from complete strangers.

The species of non-journalist newsreader became extinct in the 1980s, its demise hastened by technical advances, the transmission of live pictures through the studio and the introduction of instant reactive interviewing which demanded journalistic expertise thought to be beyond the capabilities of all but a handful of television professionals.

In reality the movement began long ago with the introduction of newscasters, a subtle but distinct difference in title. These were employed to fulfil the same functions as the old-fashioned newsreader, except that he or she was also expected to make a positive contribution to a programme by writing some of it, acting as an interviewer within it, or both. As the price for such expertise programme bosses were prepared to accept less than cut-glass accents and features which, in the case of one much-respected veteran performer, I have heard described politely as 'lived in'.

Modern presenting

A sample snapshot taken at the start of 2000 showed that of fifteen people who could be identified as fulfilling reader roles on the continuous news services, morning, lunchtime, early evening and late night national television news programmes[2] on channels available in the UK, all of them had backgrounds in journalism. Most had been reporters or correspondents, consistent with the modern policy of hiring anchors or presenters who are encouraged to play a positive role in the editorial decision-making process. Specifically, it is important they know about the stories they present, and certainly know how news is made. For the first time 'transfers' of leading players between news programmes on different channels – and sometimes between news programmes on the same channel – have taken place, with as much attendant publicity (and occasionally more acrimony) as transfers of footballers, the salaries involved subject to equal measures of envy and outrage.

The fear at one time was that as part of the justification for their pay packets the 'stars' would also be expected to make managerial decisions. Nowhere was this more apparent than in the United States, where some of the anchors wield considerable influence over programme policy, content, organization and personnel.[3] In Britain this fear has proved to be largely unfounded and the only known example of a presenter also being part of the management was, briefly, at ITN.

Without doubt news presenters have brought their experience to bear for the good of their programmes, sometimes refreshing themselves with sorties back into the field for the big occasion. On the debit side 'presenter power' can pose a genuine problem for other members of the editorial team, who may resent having their scripts changed or rewritten. Some editors are intimidated by older, very accomplished occupants of the

2. *Breakfast News*, *The One O'Clock News*, *The Six O'Clock News* and *Nine O'Clock News* (BBC 1); *Newsnight* (BBC 2); BBC News 24; BBC World Television News; *Lunchtime News*, *Early Evening News*, *Evening News* (ITV Channel 3); *Channel 4 News* (Channel 4); *Channel 5 News* (Channel 5); GMTV; Sky News.
3. See Barbara Matusow's excellent history, *The Evening Stars* (Ballentine Books, 1984).

presenter's chair, and in the face of a serious challenge to their own judgement, would probably rather back down than seek mediation from a higher editorial level, apprehensive about the choice an employer would make between a highly-paid television personality and an unknown working behind the scenes.

There also continues to be controversy over the amount of exposure news presenters should have outside their immediate sphere – whether their credibility is undermined or enhanced by their participation in quiz shows and other entertainment programmes, and how much they should engage in any debate over the news agenda. The difficulty for intelligent, active professionals is that presenting the news remains a fairly narrow occupation, and some who have done it for a while profess to hanker for a return to the good old days, when they were 'on the road'. Where the lure of being in the public eye night after night remains hard to resist it is not surprising to find presenters anxious to broaden their activities.

Behind it all, the purpose of hiring the right presenter is to help build and retain a loyal following for news and other programmes. It is significant that market research among audiences has become one of the main tools used in presenter recruitment, but it is still arguable whether viewers actually prefer one news programme and its presenting team to another or whether their choice is governed chiefly by what precedes or follows. My own view, unfashionable though it may be, is that while a hard core of viewers is devoted to each of the main news programmes, the rest can be compared with 'floating voters'. In the final analysis it is difficult to argue with those critics who suggest that who presents a news programme is of far less importance than what is shown and said on it.

Doing it in pairs

A common alternative to the solo presenter supported by the occasional reporter or specialist correspondent presenting individual items within a programme is the reading double act, first exemplified by NBC's highly successful pairing of Chet Huntley and David Brinkley, which began in the 1950s. Since then single-presenter news services all over the world have tended to follow suit whenever they have felt the need for a revamped format, and two-handed programmes have at times returned to one reader for the same reason.

The introduction of a second person into a formerly single-presenter news programme adds technical complications, possibly leading to a complete redesign of the studio set to make room for another camera. More thought has to be given to backings and lighting so there is continuity when the two presenters are shown separately in successive one-shots and then when they are seen together. Where other contributors are also expected to appear regularly there is the danger that the programme may seem to be cluttered with too many obstacles barring the way between the viewer and the news.

Most of all, it presupposes the discovery of not one but two first-class journalists willing to pool their talents for the sake of the common good. Compatibility is essential. The whole enterprise is as good as doomed, if (as has been known) the senior or more experienced member of the team is reluctant to share the studio with anyone else, regarding the new partner as an unworthy intruder, and it would be contrary to human nature if there were no pairs who simply couldn't get on.

Although the camera may show them together in the same shot for only a few seconds at either end of the programme possibly half an hour apart, the best couples are able to give the impression that, however much their individual styles differ, they hit it off

(a)

(b)

Figure 14.1 (a) Two-handed newscast. John Suchet and Kirsty Young on the ITV *Lunchtime News* set. The desk conceals computer and sound communications which are vital for the presenters. (Photo courtesy of ITN.) (b) A high-tech look and bold use of colour for the *Channel Four News* set. Considerable research goes into the look of a news set, although designers always aim to try to combine an image which conveys both authority and relaxation. (Photo Gary Braid.)

as a team, each member of it taking a genuine, continuing interest in what the other is doing, and not acting as one of two people who find themselves accidentally occupying the set at the same time. These experts somehow manage to 'bounce' off one another, each using perhaps no more than a hint of a head turn away from the partner's direction before taking up a new story. Their manner is crisp, efficient and friendly, without recourse to the cosy ultra-informality which looks and sounds so phoney.

Long discussions go on about how the work should be shared between presenters, in some cases down to the percentage of time on air an agent can negotiate for his client. What was special about the original Huntley–Brinkley partnership was that one was in Washington and the other in New York, resulting in a natural division of responsibility. Other pairs have been known to split home and foreign, 'hard' and 'soft', alternate whole items or single pages of script, or to play to the interviewing skills of one or the other. Sequences of out of vision commentaries spoken by alternate readers come across with real pace and punch, especially when the partnership consists of one female and one male.

Working with presenters

Other members of the news team are likely to regard their presenter colleagues, whether one or two, with considerable professional respect as the instruments by which a series of unrelated events become fused into a complete television programme. They know that without an anchor the transformation would be impossible. Those programmes which have a succession of reporters popping up in the studio to introduce their own items are deluding themselves if they believe they have done something different. All they have done is to share out one role between many, for the function of linking one news event to the next remains the same, whoever does it.

A few staff, perhaps other journalists among them, may have a tendency to feel resentful about the way the job of delivering the news generally rates superstar treatment, especially since presenters began to write and rewrite much of the programme. And there is no disguising the effects on programme morale as a whole if those who present the news are careless of the feelings of fellow professionals.

For the most part, the advice to give writers is to stop worrying and start appreciating how their work benefits from having it well presented. For while a poor reader can certainly ruin a good script, the corollary of that is that a good reader can improve on a poor one.

The 'writer's reader' has many qualities, mostly invisible to anyone except the insider. These include a natural gift for putting the emphasis in the right place, even on late scripts which have to be sight-read for the first time on transmission; the offer of exactly the right word or phrase which might have eluded the writer for hours; razor-sharp reactions to cues from the floor or control room, and the confidence and ability to smooth over the awkward moments which might otherwise lead to disaster.

At the other extreme is the presenter who is over-anxious, egotistical and temperamental, unwilling to accept advice, concerned only that the duties he or she is called upon to perform do not offend their public image, uncaring about the ulcer-making effects on production and editorial staff during transmission by slowing down or speeding up delivery as the mood takes them.

Not all programmes employ their presenters full time, effectively ruling out close collaboration with editorial staff, but where possible they should always be invited to attend conferences in order to absorb the feel of a day's news before they might have to read it. Proper briefing should be given for interviews and questions not left to chance.

As transmission time approaches presenters should be warned of running-order changes and awkward or unusual words. Where pronunciation dictionaries do not exist, some form of index system should be compiled as a reference. Where necessary writers should be prepared to alter words or phrases to suit the presenter's style, so long as the intended sense is not destroyed, and argue strongly when presenters take it upon themselves to make changes which would lead to inaccuracy.

Training to be a presenter

Not surprisingly, the glamour associated with presenting the news is extremely appealing to many people. Regrettably most lack the necessary experience, voice and presence. Conventional good looks by themselves are not enough. Indeed the voice matters more than appearance. Training to improve the voice is hard work and difficult whereas plenty can be done with a person's appearance. On the positive side few news organizations are complacent enough to ignore up and coming talent and many hold auditions as a matter of routine.

What they may be looking for is the occasional stand-in – someone to read at a weekend or overnight on a hot day in August, to take over temporarily while the regular stand-in replaces a presenter on special assignment or maternity. Maybe what they really want is a fresh face or voice for the traffic reports. Either way, once you have impressed them enough to earn a trial run you could be on your way.

The main difficulty lies in persuading the programme editor to put you on the audition list in the first place. Write or telephone to find the right destination for your enquiry and follow it up with a CV accompanied by a show-reel, perhaps put together on one of the commercial short courses available. But this must be relevant: if you want to be a presenter of a serious news programme then make sure you are seen presenting serious news items. Other examples of your work, paid or unpaid, with hospital or college radio and television and local radio will all help.

But remember, the chances of success at this stage are slim. As we have seen, news programmes tend to take their presenters from the ranks of experienced practitioners, and the best way of putting yourself in the shop window is first to become a journalist in television and work towards your goal gradually, picking up an all-round experience which will be invaluable if you do make it. What's more you will have a lot of fun on the way.

15

Media convergence and the future

This book is about television news, but any prospective or working journalist will probably also be using the web regularly anyway, for both research and for writing content. The Internet's ever-increasing value and profound social and economic impact have established it as the central information tool in our time. Yet television journalists, at least until early 2000, saw the web only as another information tool rather than a medium in its own right. It was only with the first big media merger – when America Online (AOL) linked with Time Warner in 2000 – that business knew for sure that the Internet would dominate home and business information and communication. Sound and vision news, movies, the web, e-commerce and e-mail would all converge. The traditional media companies like Time Warner (which included CNN) were desperate to get a foothold on the Net. The Internet companies like AOL had the opposite problem of not having much content to put onto it. Such relationships, as business consultants would say, were a perfect fit.

Television journalists should know that the Net and the web are not strictly the same thing. The World Wide Web (www) is the popular multimedia branch of the Internet. On the web, users can view not just text but also graphics, sound, video, and hyperlinks to other media or documents. As with the rest of the Internet, people can use the web to locate, read, and download documents stored on computer systems around the world. Television news can be broadcast through the web and the sound and vision quality are getting better all the time.

The Internet and television journalists

A few journalists who recognized the potential of the Internet made a lot of money on paper in the latter part of 1999 and early 2000 as their small and specialized Internet companies were bought up by the bigger companies (obviously worried that one day they would not be able to afford to buy them). It was a repeat of an earlier cycle in the business side of journalism when a few business-wise journalists back in the 1970s realized that one day many local newspapers would not be sold, but would be given away. Some of them gave up their exciting regular jobs, then went out in search of local markets in city suburbs where newspapers could be packed with advertising and then shoved through letterboxes. These newspapers have been commonplace for years. Journalists are notoriously slow at realizing the commercial possibilities of journalism, or are often indifferent to it anyway. A few of them did make big money in seeing the next hop in the way information can be a tradable consumer product.

The Internet then became an opportunity for journalism when the people who knew how to make websites work realized that the websites were only as good as the content within them. Journalists were brought in. 'Content is All' is still the maxim. Only the journalists knew about deadlines, style, layout, language and, in the case of television journalists, how to make scripting to visual images concise, attractive and effective.

During 1999–2000 there was a movement of talent as TV journalists moved onto what became known at that time as New Media. The start-up costs were low and the reward potential, and their risk, was massive. The evidence also showed that the majority of Internet companies set up actually failed within six months. Only a small minority grabbed the headlines as share prices became inflated. The Internet became a mass medium in Europe, Eastern Asia and North America by early 2000. A survey by Guardian/ICM at the beginning of 2000 showed that Internet use grew faster than any previous technological advance, including radio and television. The number of adults online in Britain grew from 29 per cent in January 1999, to 37 per cent by the end of that year.

The other aspect of that change was digital television and the rise of interactive services, the most alluring being the ability to programme a digital television to create an individual 'Me-TV' and multimedia service. If you are interested *only* in fish cookery, rugby league, soaps or just the news, then that is what you can have, all day. The technology will do the work for you. One pundit suggested that in the future people will not read books. The paradox is that one of the earliest and most successful web companies was Amazon, which sold – books.

We should not conclude that the Internet has finished changing. It is an offspring of the computer, not the traditional network of the telephone or television industry. It will evolve at the speed of the computer industry if it is to remain relevant. It is now providing services such as audio and video streams. The availability of networking along with powerful computing in portable form (laptops, two-way pagers and digital mobile telephones) is making possible an age of work-from-home communications for many news-processing operations.

Cable television

Cable television began as CATV (community antenna television) and was developed as long ago as the late 1940s to serve areas which were unable to receive television signals because of geographical difficulties or their distance from transmitters. To overcome these problems antennas were set up in areas which had good reception, and the broadcast signals distributed over a cable to subscribers. By 1995 virtually every home had access to a street cable link. That led to dozens of channels. Although the cable is on the street it remains a matter of choice whether to pay for the link to be extended into the home.

By far the most intriguing development for television news began with the introduction of 24-hour all-news channels. There are several of them around the world now, but CNN (Cable News Network) was the international cable pathfinder. CNN began broadcasting on 1 June 1980, and since then has reached far beyond its own national boundaries, transmitting its service 36 000 km (22 300 miles) into space from dish-shaped antennas in the grounds of its headquarters to satellites which send the signals back to

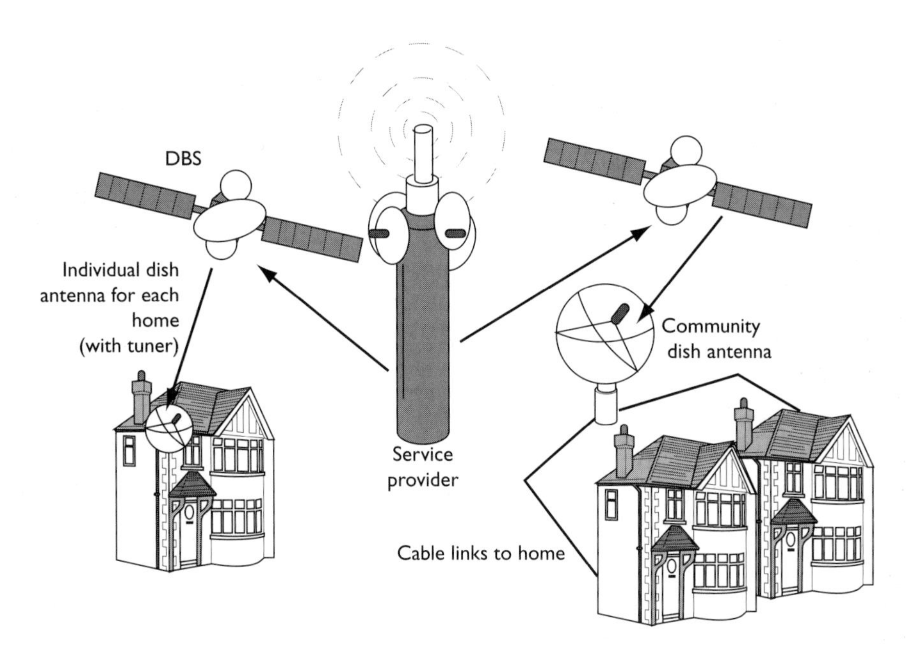

Figure 15.1 Cable and satellite. How complementary systems reach the home.

domestic and international customers on Earth. Incoming feeds from overseas bureaux are received in the same way. CNN was regarded with scorn by the big US networks when it started operations, but when the war in the Gulf started in the early 1990s it was CNN which showed everyone in news how to move fast and get stuck in. International diplomacy between Iraq, the Arab nations, the USA and the UN was quickly being conducted through the cameras of always-on-the-air CNN. CNN rivals – big news organizations like the BBC, ITN, NBC, CBS and ABC – all knew they were doing a good job in covering the Gulf War. They also privately felt that at every turn in the war they were being stuffed and outmanoeuvred by CNN, which deployed resources to live news and could put material on air instantly.

Direct satellite

Alongside cable was DBS (direct broadcasting by satellite), a system supplementing terrestrial channels with services received directly by individual households on dishes unobtrusive enough to overcome environmental and planning objections. The signals are transmitted first to high-powered satellites whose footprints (areas covered by their transmission) are the subject of international agreements. When it was launched in February 1989 from headquarters in West London, Sky included the 24-hour Sky News as one of its core services, initially beaming its wares to dish-holders via Astra, a satellite system owned by a company based in Luxembourg. The following year Sky merged with rivals British Satellite Broadcasting to become British Sky Broadcasting (BSkyB), and as part of an accelerating pattern, has long been among the many options being offered in a cable or digital package.

Figure 15.2 Interactive cable news. Viewers are offered items which they can choose to watch on another part of the screen at the same time as the 'main' broadcast.

Cable television has also led the way with interactive news, giving the viewers even greater control over what they see. A choice of complementary channels enables them to add optional services to their main picture – for example combining the news with a weather forecast, which appears in a corner of the screen. Viewers of sporting events could even choose different camera angles. Researchers at the Massachusetts Institute of Technology (MIT) in America envisage television of the future as a large flat screen on the kitchen wall showing a map of the world which yields relevant local news – textual, aural and visual – at a touch.

Teletext

Teletext, the broadcast version of videotex officially born in Britain in October 1972, has and will continue to be another source of journalistic employment, although whether those engaged on it can be accurately described as being 'in television' is a moot point. The beauty of teletext lies in its simplicity and economy. A television picture is made up of a number of horizontal lines which appear on the screen in rapid succession. No television system needs every one of those lines for carrying picture signals, and it is in a handful of the unused blank black spaces that teletext is transmitted as a series of coded electronic pulses. These can be seen as bright dots at the top of a badly adjusted television picture.

On the screen, teletext is displayed as still pages of writing and graphic diagrams, either in place of, or superimposed on, the normal pictures. Coloured headings and text, which can be made to flash for emphasis, add attraction to screen layout. Pages are strung together in the shape of 'magazines', made up of news, sport, finance, travel and weather information and so on, the viewer (reader?) selecting the required numbered page by remote control. A 'newsflash' facility, which can be chosen as a separate page, appears in a black background box cut into the conventional television picture. 'Live' subtitling of news programmes is another facility accessed by the same keypad.

The pages are compiled by journalists using all the sources available to other broadcasters and the newspapers. Copy is typed in at computer keyboards for conversion into

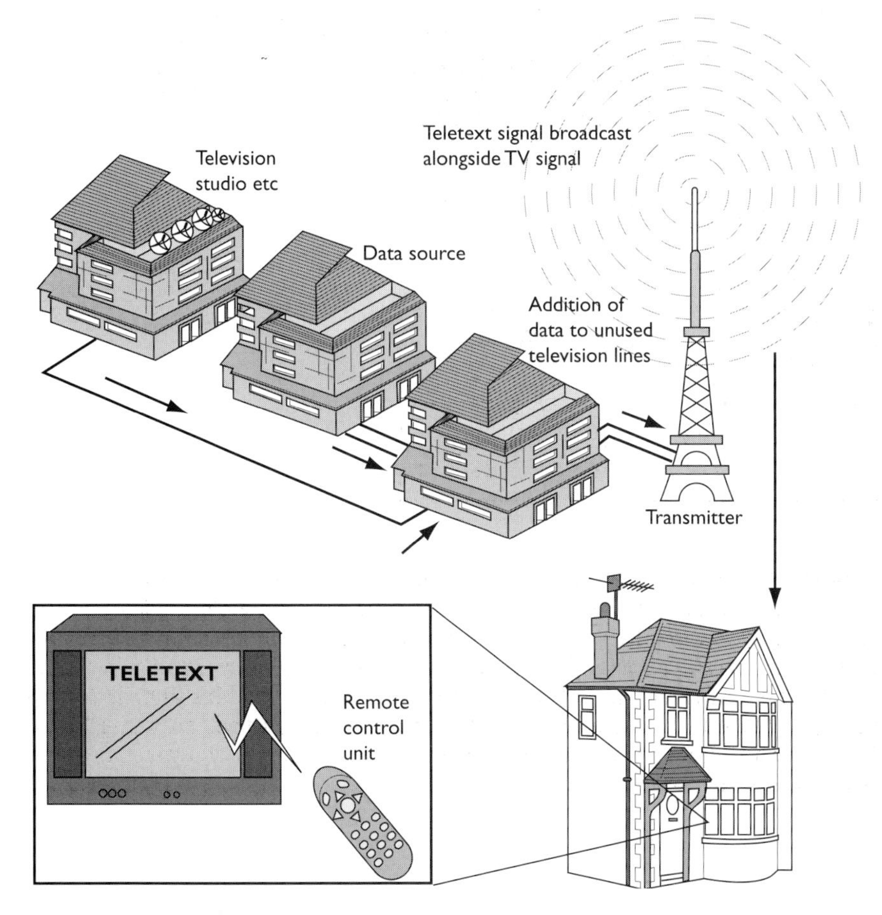

Figure 15.3 How teletext works.

the teletext signals. The speed of the system allows stories to be transmitted as quickly as they can be created.

Combination video-journalism

There is nothing very new about the idea that a journalist can also operate a camera, or operate a camera and edit the sound and vision as well. In 1990 the Canadian Broadcasting Corporation closed its regional news operation at Windsor, southern Ontario, as part of a series of economy measures, reducing a staff of more than 100 journalists, camera-operators, tape editors, producers and a support personnel to a tiny bureau. Reopening became possible only because of an agreement between CBC and the unions to use Windsor as a test bed for new technology and working practices. After a summer during which reporters were taught to shoot their own material, camera-operators were instructed in the skills of writing and reporting and picture editors were retrained to use digital technology, news programmes resumed in October 1994 – total staff: 24. Day-to-day plans and problems were handled by a joint council of management and union representatives.

CNN may lay claim to having invented the term 'video-journalist' to categorize its entry-level positions, but this facet of multi-skilling was also employed elsewhere in at least one other, narrower context. New York's first 24-hour TV news station, NY-1, gave the title to the twenty young journalists it recruited and trained more than ten years ago as camera-operator, sound recordist, picture editor and reporter combined, assigning them as district reporters with the brief to find and shoot their own stories.

Scaling down the operation to such basic levels means asking interviewees to hold their own microphones and persuading passers-by to stand in while piece-to-camera shots are framed. There are also curious contortions to be seen as reporters ask questions with one eye on the viewfinder and one ear on the camera-mounted loudspeaker, but the practitioners insist they are able to get much closer to the story than conventional crews.

Critics of NY-1 complained that the standards were low and the VJs had been introduced to save money, but the station pointed to the numbers of reporters they were able to hire compared with others. The material they produced was basic, usable quality. They were, after all, not making 'films'. They were doing news.

The television cameras available by 2000 were light, easy to operate and – if used properly – were capable of providing sound and vision of broadcast quality. A professional Betacam camera for television news can cost anything between £35 000 and £60 000. A small digital camera, with a good separate sound kit and tripod, is about £2500, although it has a shorter lifespan when worked hard and has almost zero resale value. What the lightweight digital cameras did in effect was remove the dividing line between amateur and professional. After a period of learning and experimentation with BBC News Training some experienced BBC news correspondents, already aware of filming techniques anyway, used them to compile excellent quality features when they had the time to shoot and prepare their material, aware that they would make mistakes along the way. Because it was feature or background material there was less deadline pressure. A similar service created by LWT was called The Lab. It was established so that about twenty young directors and video-journalists could train in the use of the latest equipment and pioneer the new television culture: light DV cameras used for close framing and fast-moving shots, plus editing, scripting and laying voicetracks directly into the PC. The resulting material was used in LWT's regional output and was shot and edited at a fraction of the cost of more mainstream television.

16

Conclusion

The whole emphasis of this book has deliberately been on method and technique, not motive and not the social impact of television news or its convergence with the Internet and the web. Whatever the medium of delivery – through thin widescreen TV sets with hard drives and multimedia access, or traditional receivers with terrestrial services – it is all still a matter of content which has been stroked at some stage by a human decision. Those people might be journalists, making choice on behalf of the audience. Or it might be the viewer, picking through a selection of news items he or she *chooses* to watch. As mentioned before, he or she might choose to airbrush out of their personal history events which affect the world and affect them without them realizing it, matters which they might not, at a particular moment, choose to know about.

Although it might have been tempting to suggest how individual news services should perceive their role in society, with particular emphasis on how they should approach the hypersensitive areas of news selection and processing, it would be wrong to have done so and naive to believe it would make any difference. Why a programme chooses to operate the way it does is a matter between those editorially responsible for it and the public which views it. Each news programme is guided by its own set of principles, most of which have their origins in both the prevailing attitudes of the time and the political and social ethos of the country in which it is broadcasting.

Editorial values are therefore inevitably bound to differ, and because large numbers of broadcasting organizations are owned by governments, it follows that what passes for news in one country will not necessarily match the concept of what passes for news in another. For every journalist struggling towards the goal of balanced objectivity, there are others for whom such niceties are of no concern. To put it another way, one man's objective report revealing government incompetence or corruption is another's betrayal of the society which succours him.

It is a conundrum which has existed ever since governments began to understand the power of the medium. One experienced Western journalist, seconded as news adviser to the television station of a Middle East kingdom, remembers the frustration of being quite unable to persuade those in charge that their judgement was 'at fault' because they insisted on beginning every main bulletin with news of the official activities of the King, followed by the official activities of the Queen, then the Crown Prince, and so on right the way through the royal household. All this was followed by speeches by the Prime Minister and the activities of other government ministers. Domestic news came next, most of it of stunning triviality. Then came foreign news, led by equally trivial events about countries considered friendly to the government.

Later, during the adviser's stay, a revolution took place. The monarchy was overthrown and exiled, a republic proclaimed. The editors of the news responded at once. From then on, every bulletin began with news of the official activities of the President of the Revolutionary Council, then the activities of each Council member in turn. Domestic news came next, in much the same way as before, followed by foreign news about countries which were considered friendly towards the new leadership. Former allies, together with the royal family, were completely ignored. The journalists carried on doing what they saw as their duty, serving the interests of the state. That the circumstances had changed so drastically made no difference to the principle.

My own experience, some years later, was of the assassination of a British public figure: it aroused immense interest and concern in some parts of the world, but rated no more than a passing mention by the television news service I had been asked to advise. Later discussion about the treatment of the story got nowhere: the editors were unconvinced their decision was 'wrong' in any way. They preferred to devote their limited airtime to reporting matters they considered more relevant to their audience. Who is to say they were mistaken?

Even now, in the younger democracies of Europe, it is not unknown for editors to fall into disfavour for paying insufficient attention to publicizing government 'initiatives', or too much to the opposition's suggestions. The availability of satellite-delivered international news services has greatly weakened the power and influence of state-funded services. A bunch of armed 'rebels' in a remote area of central Asia will quite happily sit on top of a mountain stronghold and watch CNN or the BBC to witness what the President of the United States has to say about the latest activities in their country. The President and his press advisers know it!

As a whole, though, the question of whether Western editorial standards have improved or dropped is entirely a matter of subjective assessment. There is a view that some broadcasters, while rightly taking more account of what their viewers are saying, have become nervous of being tarred with the brush of tabloid excess. A newscaster, unhappy with the nightly diet, has raised the concept of 'good news' reporting. Privacy, violence and intrusion have all become big issues. Internal and external 'guidelines' and 'codes of conduct' seem at times to have taken the place of individual journalistic judgement and experience.

Taste, decency and ethics

There is a valid argument that taste changes, but Decency never does. This has all to do with what social psychologists call the 'prevailing moral climate'. It depends on the time we live in and the place we live in. In most democracies that usually means these: be kind to old people, children and animals; treat people as you expect to be treated yourself; preserve religious and political freedom; maintain a free press and promote and tolerate minorities in society and support the traditional family. All news values reflect change in society. It was a big news story in 1939 when a German economist was appointed as director of a big British company, but only because everyone knew war with Germany was probable. In the 1970s the labour correspondents were busy because in Britain a monetarist Prime Minister (Margaret Thatcher) let good money drive out bad and believed business could not buck the market. That meant she would not support industries which could not survive in the market. There were strikes, job losses and closed factories. By 2000 and beyond other issues dominate social interest and as a result they have crawled up the running orders of news bulletins: education (driven by social and technological change), environment and food safety, job security and mobility, the split between people who are time rich and time poor and the constant evolution of information technology.

The climate of taste and ethics has also reminded television journalists that the fear of crime can have a worse affect on people than real crime itself. It means throwing out the old news maxim: 'If it bleeds it leads!', on the grounds that the real evidence shows that violent crime is rare; that where you live has most bearing on whether or not you are a victim of crime; that most victims of homicide are actually young children and not really drunk young men outside a bar on a Saturday night.

No sensible holder of editorial responsibility sets out to mimic the agenda of some crime-dominated newscasts elsewhere. While aberrations do occur, usually with hell to pay afterwards, mainstream television news in the United Kingdom does not, as a matter of daily routine, deliberately show gratuitous violence and gore, linger on big close-ups of the dead and dying, expose the bereaved or suffering to unnecessary intrusion, hound ordinary people for real or supposed transgressions or stuff newscasts with salacious crime stories.

By all means exercise discretion, be sure to accentuate the positive as well as the negative, refrain from the litany of routine criminal activity. But – and it is a very large but – cut out all the unpleasant bits, ignore some of the more distasteful aspects of human behaviour, avoid upsetting anyone, sanitize the news and you do the audience a greater disservice by distorting reality.

There are other concerns – for one, the definition of news. 'News is where you have a camera crew and a satellite dish' is among the more cynical, particularly abroad. That does not make that untrue. Television news cannot report a terrible war by showing pictures only of men and women in smart clothes around a table at peace talks when on the ground there is no peace. Fly a camera team to cover a story on the other side of the world, meet the expense of hotels, transport, e-mails, satellite communications, and the understandable temptation is to wring every ounce out of it, more than once, worth it or not. Another concern is the ability to make cool news judgement under pressure. If they are to exercise the delicate control necessary to produce programmes which are as balanced and thoughtful as the constraint of time allows, editors need to stand back from the fray, not allow themselves to be influenced by late incoming pictures only because they are late and available, or to succumb to the temptation of keeping reporters on the spot, long after the stories they are covering are over, just for the cosmetic effect of 'going live'.

There is also a problem faced by modern live and continuous news. There is a tyranny about live news when it deals with events which move fast and where lives are at risk. Too much analysis at too early a stage in an event can lead to false assumptions. Politicians and negotiators can now feel under pressure to act quickly rather than sit back briefly and think out the right solution. Television journalists are of course aware of all this and know the implications of demanding constant 'updates'.

There are still deficiencies in television news. Here are some.

The economy has been under-reported. Too many interviewers allow politicians to get away unchallenged with the most outrageous generalities and statements about economic affairs. Coverage of social affairs seems to be based almost entirely on stories which boil down to a plea for funds – justified or not – usually from the public purse. Sports news is reported with one hand tied behind its back: the other is bound by contractual arrangements for coverage which prevent full access to people or events. With a few honourable exceptions television reporters do not pay enough attention to the pictures: listen to their scripts. They might as well be working for radio. The same 'library' shots can be noticed cropping up, unannounced as such, to make packages more visual. There remains an obsessive love-affair with pictures of the exteriors of buildings; worse, with journalists interviewing each other. How often, it seems, is the conclusion of a substantial report followed immediately by the same reporter repeating much that was said in the preceding two or three minutes. Confusion continues to exist about the difference between information which viewers find useful and speculation – especially in 'previews' about what might happen at events yet to take place – which they do not. Alongside that comes an inability to recognize when coverage becomes repetitive and saturated. Yes, there is sometimes too much crime content. Maybe it is because the police telephone information line giving details of the latest felonies is the saviour of newsroom-based journalists with few resources struggling to fill short bulletins.

As for the value of television news as a whole – if it were possible to conduct such an experiment, it would be interesting to isolate a sample audience from any news source other than ITN, BBC, Sky News or GMTV for a few weeks and then test how much they had learned about what was happening in the world. It would be good to think they would be well-informed about a wide range of subjects. Regrettably, perhaps not.

Where does this leave the enthusiastic potential newcomer? Not too depressed, I hope. The Internet can give you what you want, but what you *want* is also not always what you *need.* For all its many inadequacies, television news remains the most reliable provider of untainted information, a worthwhile and satisfying place to be for any journalist who believes passionately in and strives unceasingly for high and impartial editorial standards. At the turn of the century the whole face of conventional television news programming has changed beyond recognition, bringing with it new and exciting possibilities as well as uncertainty over employment. For anyone wondering whether it is worth taking the risk to face such an uncertain future, there is a crumb of comfort. One sentence from *The Task of Broadcasting News*, a study made for the BBC General Advisory Council as long ago as 1976 and still relevant today, says it all: 'If ever broadcasting were pared to the proverbial bone, news would have to be that bone.'

Further reading

BBC. *Report and Accounts* (available each year in all libraries).
Bell, Martin (1995). *In Harm's Way*, Hamish Hamilton.
Cockerell, Michael (1989). *Live from Number 10: The Inside Story of Prime Ministers and Television*, Faber and Faber.
Day, Sir Robin (1993). *... But with Respect*, Weidenfeld and Nicolson.
Day, Sir Robin (1999). *Speaking For Myself*, Ebury Press.
Dougall, Robert (1973). *In and Out of the Box*, Collins Harvill.
Everton, Neil (1999). *Making Television News*, Reuters Foundation.
Hesketh, Bernard and Yorke, Ivor (1993). *An Introduction to ENG*, Focal Press.
Keith, Michael C. (1989). *Broadcast Voice Performance*, Focal Press.
Matusow, Barbara (1984). *The Evening Stars*, Ballentine Books.
Sebba, Anna (1994). *Battling for News: The Rise of the Woman Reporter*, Hodder and Stoughton.
Thompson, Chris (1994). *Non-linear Editing*, British Film Institute.
Victim Support (1992). *Victims Twice Over.*
Yorke, Ivor (1997). *Basic TV Reporting*, Focal Press.

Glossary

Note: no editorial or technical glossary can hope to be complete or wholly satisfy the expert, especially since the speed of advance in television technology has turned into a gallop. Some terms are universally understood within television news, others mean different things to different organizations or remain unique to those who apply them. In a few areas, equipment and the process of using it is known by trade name. These are among the most widely used and accepted.

ABC (1) American Broadcasting Company; (2) Australian Broadcasting Corporation.
Access time Interval between the selection of a computer function and its appearance on the screen.
Actuality See **Natural sound**.
Agency copy (wire copy) Written material received from news agencies, now usually through newsroom computer systems.
Anchor Main presenter of a news programme. See also **Newscaster**, **Newsreader**, **Presenter**.
Animation As used in television news, usually the technique of adding or changing information on a graphic.
AP Associated Press, an American news agency.
APTV Television arm of AP.
ASBU Arab States Broadcasting Union.
Aspect ratio Ratio of width to height of a television picture. Currently 4:3, but 16:9 becoming more widely used.
Assignment sheet Written instruction setting out details of an event to be covered.
Assignments desk Department responsible for assigning reporters and camera crews.
Assignments editor Executive responsible for assigning reporters and camera crews.
Aston Makers of caption-generating and other electronic equipment.
Astra Luxembourg-operated communications satellite.
Autocue Makers of electronic prompting device which enables performers to read a script while looking directly at the camera. Other makes include Autoscript, Portaprompt and Teleprompter.
Avid Manufacturer of non-linear picture editing systems. Other makers include Lightworks.
BBC British Broadcasting Corporation.
Betacam Half-inch (19 mm) video format introduced by the Sony Corporation.

Betacam SP Superior performance Betacam.

Betacart Computerized carousel system for the transmission of Beta video cassettes.

Bird Communications satellite. Named after Early Bird, the first satellite launched after the creation of Intelsat, the organization set up to establish a global system; hence 'birding' for the process of transmitting material by satellite.

BJTC Broadcast Journalism Training Council. Regulatory body for standards of broadcast journalism training in colleges and universities.

Camcorder Combined lightweight video camera and recorder.

Cans Slang for headphones.

Caption Generic term for television news artwork. See also **Graphics**.

Cassette See Video(tape) cassette.

CATV (community antenna television) System of distributing broadcast services by cable.

CBS Columbia Broadcasting System. United States radio and television network.

Ceefax BBC broadcast teletext system.

Character generator Electronic method of producing on-screen lettering in a variety of type sizes and fonts.

Chip Integrated computer circuit.

Clarke Belt Position 22 300 miles (36 000 km) above the equator in which orbiting communications satellites appear to be stationary; after the British science writer Arthur C. Clarke who first advocated the use of satellites for broadcasting.

Clean feed Actuality (natural) sound of an event free from commentary.

Closed circuit Means of distributing pictures and sound privately to selected points.

CNN (Cable News Network) All-news channel based in Atlanta, Georgia.

Colour bars A test signal in the shape of eight coloured vertical stripes.

Communications satellite Man-made device positioned in space as a means of passing television or other signals from one part of the globe to another. See also **Intelsat**.

Comsat Communications Satellite Corporation (US).

Control room/gallery Room next to or above studio from which production and technical operations are controlled during transmission of programmes.

Copy Written material for news.

Correspondent Journalist employed to report on a specialist subject or geographical location.

Countdown Time given in reverse order, usually announced aloud in the control room or by a computer transmitted voice. To ensure the smooth transition from one source to the next.

CPU (central processing unit) The computer's 'brain'.

CSO (colour separation overlay; also known as *Chromakey*) An electronic means of merging pictures from separate sources, giving the illusion, for example, that a performer in the studio is set against a pictorial background.

CU Close-up.

Cue Signal given to start or stop action.

Cut (1) An edit; (2) A deletion.

Cut-away Editing term for a shot inserted as a means of telescoping the action in a picture sequence without loss of continuity.

Cut-away/cut-in questions Questions repeated for the camera after an interview to provide a continuity bridge between edited sections.

Cut-ins Extra shots, close-ups for example, edited into the main action of a scene.

Cut-off Area of a television picture lost naturally from the domestic screen.
Cuts Also known as *trims* or *out-takes*; pictures excluded from an edited story.
Cut story Complete and edited news picture item.
Cutting/clipping Item cut or copied from a newspaper or other printed source.
Database File of information held by computer.
DBS (direct broadcasting by satellite) System of transmitting broadcast signals to individual households using high-powered satellites. Also known as *direct-to-home*.
Deaf aid Close-fitting earpiece through which a performer in the studio or in the field can be given instructions directly by editorial/production staff.
Diary story News event covered by pre-arrangement.
Dish Shaped antenna for transmitting or receiving satellite signals.
Disk Electronic storage system for storing computer information.
DOG Digitally originated graphics.
Door-stepper Informal interview obtained by waiting for the subject 'on the doorstep'.
Dope-sheet Camera-operator's detailed record of tape or film shot on location.
Down-bulletin A news item that goes in the middle or near the end of a programme.
Dry run Rehearsal without the camera.
Dub To add or re-record sound to edited pictures.
Duration Exact time/length of a programme or item within it.
DVC Digital video camera.
DVD Digital video disk.
DVE Digital video equipment.
Editor Executive in overall charge of a single news programme.
Establishing shot Scene-setting shot of people or subject.
Euronews French-based international news provider.
Eurovision European international network for the exchange of television programmes.
Eyeline The direction in which the camera sees the subject to be looking.
FCC Federal Communications Commission. The US government agency responsible for broadcasting.
Field producer Editorial supervisor of off-base assignment. See also **Fixer**.
File (1) Send a report; (2) Document stored electronically on computer.
File footage Archive/library material.
Fire brigade Editorial/camera team assigned at short notice to cover news breaks, usually abroad.
Fixer Coordinator accompanying unit in the field. Often acts as the main point of contact between home/base and team on location.
Follow-up News report based on previously broadcast or published material.
Footprint Area covered by satellite transmission.
Format (1) Overall style and 'look' of a programme; (2) Videotape size or recording pattern.
Frame A single still picture from a moving film or tape.
Frame/picture/stills store Electronic method of storing and displaying still pictures.
Free puff Slang for a news item which publicizes an event or product.
Freeze frame A single frame of video or film held to stop action.
Futures file Collection of information about items for possible future news coverage.
FX Sound effects.
Gallery See **Control room**.
Geostationary orbit Orbit in which satellites appear to remain in the same place relative to the Earth. See **Clarke Belt**.

Graphics General name for artwork or artwork department.
Gun mike See **Rifle mike**.
GV General view.
Handback/Handover Performer's form of words used to signal that his/her contribution has come to an end.
Hand-carried Equipment or material transported personally rather than sent as freight or by electronic means. See **Pigeon**.
Hand-held Camera or other equipment used without a tripod or other steadying device.
Handout Free publicity material given to news organizations.
Hand-over Form of words used as a cue for another performer (e.g. 'Now, with the sports news ...'). See **Handback**.
Hard copy Printed paper version of computer-generated material.
Hard news Straight, serious news.
Hardware Computer equipment.
HDTV High definition television. System of 1000+ lines offering superior quality pictures.
Helical scan System which scans videotape in slanting tracks.
In-cue Opening words of a news report.
Inject Live contribution to a news programme from a distant source.
Inset Visual representation of news item, usually placed over a presenter's shoulder during a newscast.
Intake/input Department responsible for news gathering. See **Assignments desk**.
Intelsat (International Telecommunications Satellite Organization) Originators of the global system by which television signals and telephony are beamed from one country to others. See **Communications satellite**.
Intro Introduction: opening sentences of a news story. Also known as *Link* or *Lead-in*.
In vision/on camera (story) Item or part item read by performer in the studio without further illustration.
ITC (Independent Television Commission) Regulatory body for independent television.
ITN (Independent Television News) Company responsible for providing national television news to several ITV companies and for services to other broadcasters.
Jump cut An ugly edit which destroys continuity, making a subject appear to jump from one position to another in successive shots.
K (kilobyte) Measurement of computer memory.
Keying colour Colour chosen to activate CSO/chromakey.
Key light Chief source of artificial light for a camera scene.
Key shot Master shot.
Lay-on Arrange coverage.
Lead (1) Opening item of a newscast; (2) Opening sentence of a broadcast news item.
Leader Portion of tape which precedes the first frame of picture, usually calibrated in seconds to aid countdown.
Lead-in See **Intro**.
Library material/tape See **File footage**.
Line Telecommunications circuit between transmitting and receiving points.
Line-up Period immediately before a recording or programme transmission during which the final technical checks are carried out.
Live As it happens.

Location Geographical position of an event.
LS Long shot.
Magazine programme Programme which is a mix of hard news and feature items.
Mic/mike Microphone.
Minicam Mobile electronic camera unit with live capability.
Modem Modulator/demodulator which allows computer signals to be transmitted by telephone.
Monitor Screen for displaying television pictures or computer-generated data.
Mono (1) Black and white (film); (2) Non-stereophonic sound.
Monopod Single extendable pole fitted to the base of a camera to keep it steady.
Multilateral Shared communications satellite booking by three or more users.
Multiplexer Vision and sound link which allows several video sources in succession to be routed at high speed on to one line for transmission.
Natural sound Sound recorded onto tape at the same time as the pictures are taken.
NBC National Broadcasting Company. US network.
Neck/personal mike Small lightweight microphone which clips onto clothing or is suspended from a cord round the neck.
Network (1) National broadcasting system; (2) Linked computer devices.
Newscaster See **Anchor**; **Newsreader**; **Presenter**.
News director (US) Executive in charge of news department.
News editor Senior journalist. In television usually concerned with news-gathering. See **Assignments desk**.
Newsreader See **Anchor**; **Newscaster**; **Presenter**.
Newswriter Newsroom-based journalist responsible for assembling and writing items for broadcast.
Noddies Reporter's simulated reaction shots for use as interview cutaways.
Non-linear editing Computerized video editing system which allows sound and pictures to be edited out of sequence.
NTSC National Television Standards Committee which gave its name to the US system of colour television.
NVQs National Vocational Qualifications.
OB Outside broadcast.
OC On camera. See **In vision**.
Onion bag String bag used for carrying videotape cassettes, so called for its resemblance to the bags in which onions are sold.
OOV (Out of vision) Commentary spoken by unseen reader. Also known as *voice-over*.
OS Outside Source. Similar to OB (above) but more common in news bulletin use.
Out-cue Final words of a news report.
Output News department responsible for the selection, processing and presentation of news material for broadcast. Counterpart of Intake/input.
Out-takes See **Cuts**.
Overlay Editing technique for matching a recorded sound track with relevant pictures. Also known as *underlay*.
PA (1) Press Association, a British domestic news agency; (2) Production assistant; (3) Programme assistant.
Package Self-contained pictorial news report comprising a number of different components.
Paintbox Electronic graphics system.
PAL (Phase Alternation (by) Line) Colour television system.

Pan Camera movement on (1) the horizontal plane; (2) the vertical plane.
PASB (programme as broadcast) Details of programme content for record and payment purposes.
PC Personal computer.
Peripheral Printer or other device linked to a computer.
Piece to camera/stand-up(per) Report spoken directly to the camera in the field.
Pigeon Traveller entrusted with the delivery of videotape from camera unit to base.
Pixel Picture element.
Presenter See **Anchor**; **Newscaster**; **Newsreader**.
Producer Person responsible for (1) entire news programme; (2) item(s) within it.
Program Set of instructions compiled to enable a computer to carry out a specific function.
Quadruplex Videotape machine with four vision heads recording across a magnetic tape 2 in (50 mm) wide.
Quantel Makers of electronic production equipment, particularly for computer graphics.
Quarter-cam Quarter-inch format video recording system.
Radio mike Microphone used with small transmitter; needs no cable link with recording equipment.
RAM (random access memory) Main computer memory. Anything put into it is lost when the machine is switched off.
Reuters British-based international agency, providing general, financial and television news services.
Reverse question See **Cut-away/cut-in questions**.
Rifle mike Directional microphone with long, barrel-like pick-up tube.
ROM (read-only memory) Program permanently built in or added to computer.
Rostrum camera Camera mounted on the photographic enlarger principle to control taping of maps and other static objects.
Rough cut First assembly of tape edited to its approximately pre-selected order and duration.
Running order/rundown Order of transmission of items in a programme.
Run through Rehearsal.
Run up The time considered necessary for technical equipment to become fully operational.
Rushes Unedited raw material from the camera.
RX Recording.
Scanner Mobile control centre serving outside broadcast unit.
SECAM (Sequence Couleur Avec Memoire) French colour television system.
Shot-list Detailed description of each scene in edited tape or film, from which the commentary is written to match the pictures.
Skillset London-based broadcast, film and video industry training organization.
Sky News BSkyB 24-hour news channel.
SOC (standard outcue) Standard phrase spoken by a programme's team of reporters at the end of every contribution.
Soft (1) A shot that is slightly out of focus; (2) Opposite of hard news.
Software Computer programs.
SOT Sound on tape.
Soundbite Interview or speech extract chosen for inclusion in edited news package.
Sound track(s) Area of tape on which sound is recorded.
SOVT Sound on VT.

Split screen Picture composed of two separate elements, each occupying half of the screen area. (A picture with more than two elements is known as a multi-screen.)
Stick mike Stick-shaped microphone much favoured for news work for speed of preparation and ease of use.
Still A single picture.
Still frame See **Freeze frame**.
Stock Raw unused tape.
Stringer A freelance contributor employed on a regular basis.
Studio spot (Usually) a contribution made live in a studio by a journalist other than the main presenter(s).
Superimposition Usually abbreviated to super or more commonly now: CapGen. Electronic combination of two or more pictures to give extra information on the screen (often a speaker's name or title).
Talk-back One-way sound link between the control room and other technical areas.
Talking head Any interviewee; also used pejoratively in the sense that to have too many talking heads on a news programme is considered unimaginative.
Teleprompter See **Autocue**.
Teletext Broadcast videotex. On-screen text information transmitted on unused lines within the television signal.
Terminal Computer keyboard. See also **VTD/VDU**.
Tilt Vertical panning movement of the camera.
Transponder (Transmitter/responder) On-board satellite equipment which receives and passes on a telecommunications signal.
Tripod Adjustable three-legged stand fixed to the base of a camera to keep it steady.
TVRO Television receive only.
Two-shot A shot of two people.
TX Transmission.
Unilateral Exclusive use by one broadcasting organization of communications satellite or other links.
Upcut US term for the accidental overlapping of two sound sources (e.g. live commentary running into recorded sound).
VDT/VDU Visual display terminal/unit. Display screen linked to computer.
Videotape (VT/VTR) System of recording television pictures and sound onto magnetic tape.
Video(tape) cassette Container which allows tape to be threaded automatically into cameras and recorders.
Videotex 'Written' information distributed to television/display screens from central computers.
Viewdata Non-broadcast videotex accessed over the telephone.
Vision story See **In vision/on camera (story)**.
VNR (video news release) Video version of written press release.
Voice-over See **OOV**.
Vox pop (Vox populi) A series of usually very short interviews on a specific topic, often with people selected at random, and edited together to give a cross-section of opinion.
Whip pan (zip pan) Very high speed panning movement of the camera.
Wild track/wild sound Recorded sound which is related to but not synchronized with the picture.
Wipe (1) An electronic production technique akin to 'turning the page'; (2) Erase.
www World Wide Web.

WYSIWYG (whizzywig) Literally: What You See Is What You Get! Warning to journalists that what they type into a computer is what gets onto the broadcasting screen. The computer cannot be blamed because it is only a tool for the human hand to deploy.

Zip Moving scrolling words on the screen, usually at the bottom.

Zoom lens A lens giving a variable focal length.

Index

http://www.focalpress.com

Visit our web site for:

- The latest information on new and forthcoming Focal Press titles
- Technical articles from industry experts
- Special offers
- Our email news service

Join our Focal Press Bookbuyers' Club

As a member, you will enjoy the following benefits:

- Special discounts on new and best-selling titles
- Advance information on forthcoming Focal Press books
- A quarterly newsletter highlighting special offers
- A 30-day guarantee on purchased titles

Membership is FREE. To join, supply your name, company, address, phone/fax numbers and email address to:

USA
Christine Degon, Product Manager
Email: christine.degon@bhusa.com
Fax: +1 781 904 2620
Address: Focal Press,
225 Wildwood Ave, Woburn,
MA 01801, USA

Europe and rest of World
Elaine Hill, Promotions Controller
Email: elaine.hill@repp.co.uk
Fax: +44 (0)1865 314572
Address: Focal Press, Linacre House,
Jordan Hill, Oxford,
UK, OX2 8DP

Catalogue

For information on all Focal Press titles, we will be happy to send you a free copy of the Focal Press catalogue:

USA
Email: christine.degon@bhusa.com

Europe and rest of World
Email: carol.burgess@repp.co.uk
Tel: +44 (0)1865 314693

Potential authors

If you have an idea for a book, please get in touch:

USA
Terri Jadick, Associate Editor
Email: terri.jadick@bhusa.com
Tel: +1 781 904 2646
Fax: +1 781 904 2640

Europe and rest of World
Christina Donaldson, Editorial Assistant
Email: christina.donaldson@repp.co.uk
Tel: +44 (0)1865 314027
Fax: +44 (0)1865 314572